# Public-Private Partnerships for Urban Water Utilities

TRENDS AND POLICY OPTIONS No. 8

# Public-Private Partnerships for Urban Water Utilities

## A Review of Experiences in Developing Countries

Philippe Marin

1818 H Street NW
Washington DC 20433
Telephone: 202-473-1000
Internet: www.worldbank.org
E-mail: feedback@worldbank.org



1 2 3 4 12 11 10 09



ISBN 978-0-8213-7956-1
eISBN: 978-0-8213-7957-8
DOI: 10.1596/978-0-8213-7956-1

**Library of Congress Cataloging-in-Publication Data**
Marin, Philippe, 1965-
Public-private partnerships for urban water utilities : a review of experiences in developing countries / Philippe Marin.
p. cm.
"February 2009."
ISBN 978-0-8213-7956-1 (pbk.) -- ISBN 978-0-8213-7957-8 (e-book)
1. Water utilities—Developing countries. 2. Public-private sector cooperation—Developing countries. I. Title.
HD4465.D44M37 2009
363.6'1091724--dc22

2009019190

Cover: Naylor Design, Inc.

# CONTENTS

## FIGURES

## TABLES

# FOREWORD

In its look back at more than 15 years of experience with public-private partnerships (PPPs) for urban water utilities in developing countries, this book examines the performance of a large sample of PPPs in different regions. Directed at policy makers in governments as well as donors and other stakeholders, its aim is to better understand the contribution of water PPPs to help improve the provision of water and sanitation services to the urban populations of the developing world.

This report shows that despite difficulties in several countries, water PPP has largely passed the test of time. The urban population served by private water operators in the developing world has been growing every year since 1990, reaching about 160 million people by 2007. The positive record on service and efficiency improvements reaffirms the value of PPPs, even though the level of private financing did not match initial expectations. Over time, a more realistic market has developed, with a growing number of private investors from developing countries and with contract designs based on a more pragmatic allocation of risks between partners. What emerges from examining the available empirical evidence is that well-designed partnerships between the public and the private sectors are a valid option to turn around poorly performing water utilities in developing countries.

The water sector has many features that set it apart from other infrastructure sectors. This book suggests that a careful consideration of these specificities is important for successfully involving private operators. In the challenging environment of many developing countries, the main focus of water PPP should not be about attracting direct private investment, but rather about using private operators to improve service quality and efficiency. This approach fosters a virtuous circle whereby the utility improves its financial situation and gradually becomes able to finance a larger share of its investment needs. Although concessions have worked in a few places, contractual arrangements that combine private operation with public financing of investment appear to be the most sustainable option in many countries. An obvious implication for governments and donors is that they need to remain heavily engaged in the water sector, especially in the poorest countries.

These findings are important. When money—public and private—is scarce, improvements in service and efficiency are essential, leading to a better functioning of water utilities and eventually to improved creditworthiness, which is even more true at a time of global financial crisis. Well-designed public-private partnerships can help. Decision makers in developing countries need to have various options to tackle the many challenges of water utilities. This report confirms that PPPs can be one of them.

Jamal Saghir
Director
Energy, Transport, and Water
Chair, Water Sector Board
World Bank

Zoubida Allaoua
Director
Finance, Economics, and Urban Department
World Bank
Chair, PPIAF Program Council

# ACKNOWLEDGMENTS

This report presents the key findings from a study of experiences with public-private partnerships for water supply in developing countries. The study was carried out by the Water Anchor of the Energy, Transport, and Water department (ETWWA) of the World Bank, in partnership with the Public-Private Infrastructure Advisory Facility (PPIAF), between May 2006 and June 2008.

This study was led by Philippe Marin, who is also the main author of this report. The study team included Luis Andrés (LCSSD), Alexander Danilenko (Water Supply and Sanitation Program), Bertrand Dardenne (consultant), Matar Fall (ETWWA), Jonathan Halpern (ETWWA), Ada Karina Izaguirre (Finance, Economics, and Urban department), Alain Locussol (consultant), and Josses Mugabi (consultant). The study was supervised by Abel Mejia (ETWWA). Bertrand Dardenne, Jonathan Halpern, Ada Karina Izaguirre, and Josses Mugabi collaborated in the drafting of some portions of the report. Special thanks are due to Jyoti Shukla (PPIAF) and Clemencia Torres de Mästle (PPIAF) for continuous support during the implementation of the study.

Many consultants participated actively in the collection and analysis of the data covered in the study. They include Bertrand Dardenne, Jorge Ducci, Hazim El-Nasser, Jean Pierre Florentin, Mauricio Fourniol, Angela Gonzalez, Alain Locussol, Jean Pierre Mas, Josses Mugabi, Silver Mugisha, William Muhairwe, Mariles Navarro, Ian Palmer, Gabriela Prunier, Julio Miguel Silva, Alejandro Valencia, Richard Verspyck, and Guillermo Yepes.

Many colleagues in the World Bank Group helped in data gathering and contributed valuable feedback and comments. They include Thadeu Abicalil, Oscar Alvarado, Aldo Baietti, Alexander Bakalian, Sabine Beddies, Ventura Bengoechea, Lorenzo Bertolini, Franck Bousquet, Greg

Bowder, Xavier Chauvot de Beauchêne, Jeffrey Delmon, Katharina Gassner, Philippe Huc, Vijay Jagannathan, Jan Janssens, Suhail Jme'An, Jonathan Kamkwalala, Bill Kingdom, Peter Kolsky, James Leigland, Patricia López, Midori Makino, Cledan Mandri-Perrott, Seema Manghee, Pier Mantovani, Manuel Mariño, Alexander McPhail, Iain Menzies, Eustache Ouayoro, Nataliya Pushak, Catherine Revels, Gustavo Saltiel, Manuel Schiffler, Jordan Schwartz, Avjeet Singh, David Sislen, Mario Suardi, Luiz Tavares, Carolinez van den Berg, Meike van Ginneken, Patricia Veevers-Carter, Carlos Velez, Jane Walker, and Michael Webster. Special thanks are due to Luis Andrés, Katharina Gassner, and the team from the International Benchmarking Network for Water and Sanitation Utilities (IBNET) for kindly facilitating access to their data set.

The project team would also like to thank the many representatives of governments, regulatory agencies, and the private sector who provided data and information for the study. Special thanks are due to Richard Franceys (Cranfield University), José Luis Guasch (World Bank), Felipe Medalla (University of the Philippines), Fadel N'Daw (Millenium Drinking Water and Drainage [PEPAM], Senegal), Gerard Payen (Aquafed), Paul Reiter (International Water Association), and Robin Simpson (Consumers International) for providing valuable feedback and comments to the final draft of the report. Special thanks to my colleagues Janique Racine in PPIAF; Steve Kennedy; and Richard Crabbe, Andrés Meneses, and Janice Tuten in the World Bank's Office of the Publisher.

# ABOUT THE AUTHOR

**Philippe Marin** is a senior water and sanitation specialist in the Energy, Transport, and Water department of the Sustainable Development Network of the World Bank. An expert on water utilities reforms and public-private partnerships, he has worked in more than 40 developing and developed countries on institutional reforms, infrastructure finance, and utilities management. He joined the World Bank Group in 2001 and has more than 15 years of experience in the water sector, gained with the private sector and several international financial institutions. He holds an MSc degree in engineering from Institut National Agronomique Paris-Grignon and an MBA degree from INSEAD in Fontainebleau, France.

# ABBREVIATIONS

| | |
|---|---|
| BOT | build, operate, and transfer |
| ETWWA | Energy, Transport, and Water department (of the World Bank) |
| GPOBA | Global Partnership for Output-Based Aid |
| IBNET | International Benchmarking Network for Water and Sanitation Utilities |
| NRW | non-revenue water |
| O&M | operation and maintenance |
| PME | Programa de Modernización de Empresas (Colombia) |
| PPI | Private Participation in Infrastructure (World Bank/PPIAF Projects Database) |
| PPIAF | Public-Private Infrastructure Advisory Facility |
| PPP | public-private partnership |

# OVERVIEW

Many governments in the 1990s embarked on ambitious reforms of their urban water supply and sanitation (WSS) services that often included delegating the management of utilities to private operators under various contractual arrangements. Hopes were high that public-private partnerships (PPPs) would turn around poorly performing public utilities by bringing new expertise, financial resources, and a more commercial orientation. Since 1990, more than 260 contracts have been awarded to private operators for the management of urban water and sanitation utilities in the developing world.

PPP projects in the water sector have been controversial, particularly after a series of highly publicized contract terminations in recent years raised doubts about the suitability of the approach for developing countries. The lack of data on the populations served and on the quality of services provided has made it difficult to assess the overall contribution of PPP projects in developing countries. The debate has sometimes been driven more by ideology than by objective results, and the performance record of many PPP projects has never been scrutinized. Today, about 7 percent of the urban population in the developing world is served by private operators, and doubts remain over the merits of this approach for improving the performance of water utilities in developing and transition countries.

This study seeks to contribute to the understanding of the performance of PPP projects in urban water utilities in developing countries. It focuses on projects in which a private operator is introduced to run the utility, consequently excluding build, operate, and transfer (BOT) projects and similar arrangements limited to the construction and operation of treatment facilities. It reviews the overall spread of urban water PPPs during the past 15 years and seeks to respond to the questions of whether and how they have helped to improve services and to expand access for the populations concerned. The study analyzes performance data from more than 65 large water PPP projects that have been in place for at least five years (three years in the case of management contracts), a sample that represents a combined population of about 100 million people—close to half of the urban population that has been served by private water operators some time between 1990 and 2007. This sample represents, by size of population served, close to 80 percent of the water PPP projects that were awarded before 2003 and that have been active for at least three years.

Four dimensions of performance are analyzed: access (coverage expansion), quality of service, operational efficiency, and tariff levels. The analysis focuses on the net improvements and actual impact for the concerned populations, rather than whether contractual targets were met. Based on what worked and what did not, conclusions are then drawn on how governments can better harness private initiative to improve water supply and sanitation services in the developing world.

## Growth of Water PPPs since 1990

Between 1991 and 2000, the population served by private operators in developing and transition countries grew steadily from 6 million to 94 million. The number of developing and transition countries with active water PPP projects increased from 4 to 38. However, problems started to appear in the late 1990s, and the number of new PPP contract awards began to decrease.

Although the general perception is that water PPPs in developing countries are on the decline, the situation is more nuanced. The population served by private water operators in developing and emerging countries has continued to increase steadily, from 94 million in 2000 to more than 160 million by the end of 2007. Large countries such as Algeria, China, Malaysia, and the Russian Federation have started to rely on private water operators on a large scale. Out of the more than 260 contracts awarded since 1990, 84 percent were still active at the end of 2007, and only 9 percent had been terminated early. Most cancellations were in Sub-Saharan Africa, a challenging region for reform, and in Latin America, among concession schemes.

## Performance of Water PPP Projects

This study analyzes four dimensions of performance: access (coverage expansion), quality of service, operational efficiency, and tariff levels.

*Access*

The analysis of the impact of PPPs on access to piped water focuses on concessions (where most of the investment is funded by the private partner) and leases-affermages (where it is mostly funded by the public partner). Overall, it is estimated that water PPP projects have provided access to piped water for more than 24 million people in developing countries since 1990.

The overall performance of concessions for expanding access to service has been mixed. The 30 large concessions under review provided access to piped water for about 17 million people, but many of those concessions failed to invest the amount of private funding they had originally committed, and did not always meet their original contractual targets for coverage. Many of the good performers were concessions in which private financing was actually complemented by public funding (Colombia, Guayaquil in Ecuador, and Cordoba in Argentina).

The performance of leases-affermages was usually more satisfactory. In Sub-Saharan Africa, the affermage approach, with investment carried out by a public asset-holding company, has been very successful for expanding access in Senegal. The case of Côte d'Ivoire deserves special note: almost 3 million people there have gained access to piped water through household connections since 1990—entirely financed through cash-flow generation from tariff revenues without any government money.

*Quality of Service*

Often water PPPs have substantially improved service quality, especially by reducing water rationing. Rationing is possibly the number one quality challenge for many water utilities in the developing world. Without service continuity, meeting drinking water standards cannot be guaranteed because of the risk of infiltration in pipes. The poor, who often live at the low-pressure ends of distribution networks and cannot afford coping equipment (such as private wells, roof tanks, and filters), are disproportionately affected. Once water rationing becomes the standard practice in a utility, it is very hard to reverse. Frequent surges in pressure speed up the deterioration of the network, and any attempt to increase the average service pressure causes more burst pipes and lost water. In this context, it is remarkable that many of the PPPs that started from a situation of water rationing succeeded in improving service continuity and that some even managed to reestablish continuous service.

A good illustration is provided by the case in Colombia, where private operators have consistently succeeded in improving service continuity in many cities and towns, often starting from highly deteriorated systems. Private operators also have a good track record of reducing water rationing in Western Africa (Guinea, Gabon, Niger, and Senegal). Several management contracts also achieved notable progress despite their short duration. However, not all PPPs have succeeded in improving service continuity. For instance, in Manila (the Philippines) the concessionaire in the Western zone failed while that in the Eastern zone succeeded.

### *Operational Efficiency*

A key objective for incorporating private operators is to improve operating efficiency. Although utility operation has multiple facets, in practice, the overall efficiency of an operator can be broadly captured by three main indicators: water losses, bill collection, and labor productivity.

***Water Losses.*** Controlling water losses is a priority for any well-run utility. Recent multicountry studies by Andrés, Guasch, Haven, and Foster (2008) and Gassner, Popov, and Pushak (2008b) found that private operators were effective in reducing water losses. Confirming their findings, this study found that many private operators succeeded in reducing water losses, notably in Western Africa, Brazil, Colombia, Morocco, and Eastern Manila in the Philippines. In some cases, private operators even reduced nonrevenue water (NRW) to less than 15 percent, a rate similar to that in some of the best-performing utilities in developed countries.

However, not all the PPP projects reviewed achieved a significant reduction in water losses. For instance, in Guayaquil (Ecuador), Maputo (Mozambique), and Western Manila (the Philippines), no notable progress was achieved, and NRW remained at very high levels (more than 50 percent). In several countries, including Argentina, tracking the actual evolution of water losses is difficult because a large proportion of residential customers are billed on estimated, not actual, consumption. Less than half of the management contracts under review showed sizable progress.

***Bill Collection.*** It is common for poorly performing public utilities to have low bill-collection rates because of lax enforcement and the fact that people often resent paying for poor services. Bill collection is an area in which it is widely assumed that private operators are efficient, because of direct financial incentives. Indeed, this study found that, in most cases, the introduction of a private operator markedly improved collection rates. This is the dimension in which the positive contribution of management contracts was most consistent, with all the projects in the sample achieving significant improvements.

***Labor Productivity.*** There is strong evidence that the introduction of private operators resulted in improvements in labor productivity (measured as the number of staff per thousand customers), achieved through both staffing reductions and increases in the customer base. Many of the utilities concerned were overstaffed, and PPP projects were often accompanied by significant layoffs, ranging from 20 percent to 65 percent of the initial labor force. The layoffs were often motivated not just by overstaffing but also by the need to change the overall profile of employees and to hire more skilled staff.

***Overall Efficiency.*** When analyzing these three indicators of performance in combination, operational efficiency appears to be the area in which the positive contribution of private operators has been the most consistent. To capture the full impact of a PPP on operational efficiency, detailed financial analysis of individual projects would have been required, going beyond the scope of this study. Nonetheless, some general conclusions can be drawn.

The overall efficiency of *concessionaires* is hard to judge, because they are responsible for both operations and investment; their investment efficiency was not addressed in this study. In the case of Manila, a detailed analysis by the regulator showed that the concessionaire in the Eastern zone had significantly improved operational efficiency, while the one in the Western zone had not. In the case of Argentina, it is not clear whether concessionaires achieved much improvement in efficiency.

In *leases-affermages*, the efficiency of private operators is easier to assess, because the responsibility for operation and that for investment are separated between the private and public partners. Detailed information available for such projects in Senegal and Cartagena (Colombia) showed that clear gains in operational efficiency were achieved, which were passed to consumers over time through tariff reductions in real terms.

*Management contracts* entail only a limited transfer of responsibility to private operators, giving them limited control over a utility's labor force. Efficiency improvements under management contracts—measured using the global efficiency index (the ratio of water billed and paid for to water produced, a measure that combines water loss reduction and improved bill collection)—were significant in most cases under review.

### *Tariff Levels*

Most poorly performing public utilities in developing countries have water tariffs that are well below cost-recovery levels, and raising them is often a necessary component of reform toward financial sustainability. In practice, the potential impact of a PPP on the tariff depends on how far the initial tariff level is from the cost-recovery level and on the extent of efficiency gains that can be made by the private operator—two factors that move in opposite directions and can be of very large magnitude in developing countries.

The evolution of tariff levels in a number of PPP projects was analyzed as part of the present study. In most cases, tariffs rose over time, but the underlying reasons, as well as whether those increases were justified, could not be assessed. Analyzing the impact of PPPs on tariffs can be easily misleading, because it is heavily dependent on prevailing tariff policies. Tariff increases are not necessarily a bad thing for customers when they also translate into wider access to better services, as happened under many PPPs. In many developing countries, low water tariffs mostly benefit the connected middle class and work against the interests of the unconnected urban poor who need to access water from often unsafe and/or more expensive sources. It is likely that many of the poor households that gained access to piped water under PPP projects ended up paying a lower price for water than when they were not connected to the network. It must also be noted that in a few recorded cases, private operators made large enough efficiency gains to allow for significant tariff reductions in real terms after a few years.

The evidence from the literature on the impact of PPPs on tariffs is also largely inconclusive. Costs are greatly affected by local factors, such as raw water availability, and comparing tariff levels between private and public utilities can be misleading because of differences in the legal, administrative, and financial frameworks in which the two sets of utilities operate. The recent study by Gassner, Popov, and Pushak (2008a) uses a large sample to control for the many exogenous factors. It found no statistically significant difference in water tariffs between comparable public and private utilities.

## Key Findings

### *Water PPPs Are a Viable Option in Developing Countries*

Despite limitations related to data accessibility and reliability, and the ambiguity of indicators, the analysis of the four dimensions of performance (access, quality of service, operational efficiency, and tariff levels) suggests that the overall performance of water PPP projects has been generally quite satisfactory. Several PPP projects performed well on coverage (access), service quality, and efficiency together. More performed well in one or two key aspects. Some brought sizable improvements to the populations they served even though they proved unsustainable and were terminated early. A few others failed to achieve any meaningful results by most accounts.

It is noteworthy that out of 65 developing countries that embarked on water PPPs during the past two decades, at least 41 still had private water operators, and 84 percent of all awarded contracts were still active, by the end of 2007. Twenty-four countries had reverted to public management, and several contracts had been terminated early following conflicts between the parties. These numbers are not unreasonable considering what has in practice been a market test of a wide variety of contractual designs in many different (and often challenging) environments. Details do matter; the choice of contractual designs, as well as the willingness of the public and private partners to make it work during implementation, have proved to be major determinants in the final outcome.

To draw a general picture of the overall outcome of water PPP projects in the developing world, this study attempted a broad classification. A total of 205 million people in developing and emerging countries have been served by water PPP projects at some point during the past 15 years. Of these, 160 million people were still being served at the end of 2007, while about 45 million people had been served by PPPs that were either terminated early or not renewed at expiration.

Among the 160 million people served by private operators in 2007, about 50 million are served by PPP projects that can be classified as broadly successful. These are projects that have brought significant benefits to the population and where a working relationship has developed over time between the public and private partners. Successful PPP projects exist in all regions of the developing world, including Latin America (Colombia, Chile, Guayaquil in Ecuador, and several concessions in Brazil and Argentina), Sub-Saharan Africa (Côte d'Ivoire, Gabon, and Senegal), Asia (Eastern Manila in the Philippines), Eastern Europe and Central Asia (Yerevan in Armenia), and the Middle East and North Africa (Morocco). Active PPP projects whose performance was mixed or disappointing are estimated to represent a population of about 20 million. The remainder (90 million people) receive service under PPP projects that were not reviewed in this study, most of these projects being recent (awarded since 2003).

### *The Most Consistent Contribution of Private Operators Has Been Improved Efficiency*

In the 1990s, the main attraction of PPPs in the sector was their supposed ability to supply private finance. Experience has shown that this was largely the wrong focus. The review of the cases that worked shows that the biggest contribution that private operators can make is improving operational efficiency and service quality. These improvements have a major impact on access to financing, but indirectly. Customers become more willing to pay their bills when service improves and more efficient operation creates more cash flow from operations to invest in expansion, which in turn increases the customer base and revenues. As creditworthiness improves, a utility can more easily access funding and invest in service expansion. An efficient operator will make good use of the funding that is available for investment, regardless of whether the funding comes from public or private sources.

*Contractual Arrangements for Water PPPs Have Evolved Differently in Different Regions*
A large proportion of the PPPs that were awarded during the 1990s, particularly in Latin America, focused on attracting private funding and therefore adopted the concession scheme. The early termination of many of these concessions demonstrated the inherent vulnerability of this approach in the volatile economic environment of developing countries. Colombia was the first to depart from the standard concession approach, using the mixed-ownership companies approach or providing public grants to private concessionaires to accelerate investment. Many of these hybrid PPPs had positive results. In other regions, several countries experimented with long-term PPPs that combined private operation with public investment, such as leases-affermages, mixed-ownership companies, and management contracts.

## Looking Forward

The findings of this study suggest that a new approach is emerging for maximizing the potential contribution of private water operators in the developing world. The focus of PPP should be on using private operators to improve operational efficiency and quality of service, instead of primarily trying to attract private financing. A new generation of water PPP projects already has been gradually emerging, as these elements were being internalized by the market. In practice, the optimal modalities for financing investments depend on the specific situation of each country.

*Emerging Options to Finance Long-Term PPPs: Toward Hybrid Models*
Despite the difficulties that were experienced with concessions in several countries, private financing should not be discarded altogether. It has started to prove viable in a few of the more advanced developing countries, where medium- and long-term private debt in local currency have become available. However, in most of the developing world, the bulk of the large capital outlays required to expand access in the near future will have to come from public sources.

More and more countries are adopting a PPP model in which investment is largely funded by public money, with the private operator focusing on improving service and operational efficiency. In practice, funding for investment under these mixed-financing PPP projects comes from a combination of direct cash flows from revenues, with a variable mix of government and private sources that tends to make the traditional dichotomy between leases-affermages and concessions increasingly obsolete. Several successful approaches have been developed over the past decade:

- Concessions that rely largely on revenue cash flow for investment, with cross-subsidies from electricity sales (Gabon), tariff surcharges (Côte d'Ivoire), or both (Morocco).
- Affermages, as developed in Western Africa, bolstered by enhanced incentives for operational efficiency, a program of subsidized connections to expand coverage for the poor, and a gradual move to full cost recovery through tariffs (Senegal, Niger, and now Cameroon).
- Mixed-ownership companies, as used in Latin America (Colombia, La Havana in Cuba, and Saltillo in Mexico) and several countries of Eastern Europe (the Czech Republic and Hungary).
- Concessions with public grants for investments to spearhead access expansion or rehabilitation while minimizing the impact on tariffs. This is typified by the PPPs in Colombia designed under that country's *Programa de Modernización de Empresas* (PME); a similar approach has been adopted in Guayaquil in Ecuador and in a few concessions in Argentina (Cordoba and Salta).

***New Private Water Operators from Developing Countries***

In parallel with a shift in PPP models, many new players have been entering the market. In 2000, five international water companies accounted for about 80 percent of the water PPP market in developing countries. Since 2001, private operators from developing countries have signed most of the new contracts, and some international operators have also transferred their existing contracts to local investors.

Some 90 percent of the growth in the number of people served by PPP projects since 2001 is due to private operators from developing countries. By 2007, local private water operators served more than 67 million people, or more than 40 percent of the market. This study identified as many as 28 large private operators from developing and emerging countries, each serving a combined population of at least 400,000 people. In East Manila in the Philippines as well as several PPP projects in Argentina, Brazil, and Colombia, local private investors have proved their capacity to learn the trade, deliver good performance, and become credible players.

It would be hard to overestimate the importance of this new trend. Not only do these new operators provide much-needed competition in the sector, but they also may have a better capacity to manage the various risks inherent in the urban water utility business. Their understanding of local culture can allow them to more easily establish a viable partnership with local authorities and better mitigate political risks. They are also probably better suited than their international competitors to serving small cities and towns, where the needs are considerable.

## Toward a More Balanced Debate

It is clear from the many experiences of the past 15 years that public-private partnership is not a magic formula to address all the multiple issues of failing public water utilities in the developing world. For many governments in developing and transition countries, PPP projects have proved to be complex undertakings that carry strong political risks and large uncertainties as to the magnitude and timing of the expected benefits. Contractual targets are difficult to set and baseline data are seldom reliable; they generate many opportunities for conflict. Private operators do not always deliver and have a tendency to seek renegotiations to their advantage. Reforms can become easily subverted by vested interests. Many obstacles can lead to conflicts and costly early termination. Still, the overall performance of water PPPs is more positive than is commonly believed. PPP projects for urban water utilities have brought significant benefits to tens of millions of people in the developing world.

Transferring a majority of urban water services to private operators is unlikely to be the chosen option for most developing countries. But having a few water supply PPPs in a country can still be very beneficial, by generating much-needed pressures to move the whole sector toward higher levels of performance. The public water utilities that have succeeded in improving performance are those that have applied sound commercial management principles, emphasizing financial viability, accountability, and customer service. Complacency is the worst enemy of public utilities, and it is rooted in the assumption that poor service has no consequences. That attitude makes it difficult for even the most skilled and best-intentioned public managers to introduce and sustain improvements in the face of the various groups that have vested interests in the status quo. In that sense, the actual contribution of water PPPs may be greater than that achieved in specific projects—through the introduction of a much-needed sense of competition and accountability in an erstwhile monopolistic sector.

Many public water supply utilities in the developing world are also opening the door to the private sector through practices that fall short of delegated management but open the way for a new, broader approach for private sector involvement. These include other forms of providing operational expertise, such as performance-based service contracts, twinning, and subcontracting. The private sector is also gaining a new role with public utilities thanks to the increasing recourse to private financiers in the most advanced countries—with, in addition to nonrecourse BOT projects for treatment facilities, the recent development of subsovereign borrowing or sale of minority equity shares through initial public offerings (IPOs). Finally, publicly owned water utilities are starting to look for delegated management or other contracts outside of their jurisdiction, where they contractually become private partners. All this makes the traditional boundaries between public and private water utilities increasingly blurred, fostering a more buoyant and competitive market and more choices for decision makers in government.

The private sector has much to offer, and in many forms. To tackle the immense challenges facing the urban water sector in developing countries, policy makers need all the help they can get. It might just be time for a broader concept of partnership, one that includes all and excludes none.

# 1. INTRODUCTION

In the 1990s, many governments embarked on a series of reforms of urban water supply and sanitation (WSS) services, often with support from international financial institutions. Reforms were badly needed: millions of people lacked access to piped water and sanitation services; and for millions of others, service was often poor. Deteriorated infrastructure, fast urban growth, and large investment needs coexisted with poorly run utilities, artificially low tariffs, and scarce fiscal resources. Efforts to strengthen publicly managed utilities had proved largely incapable of addressing the sector's mounting challenges.

A major component of the new reforms was a heavy reliance on the private sector. For governments that lacked enough fiscal resources to cover the financial losses of public utilities and to invest in infrastructure rehabilitation and expansion, public-private partnerships (PPPs) for water utilities seemed to be an attractive solution. Hopes were high that with their expertise and financial resources, private operators would provide better services for a larger number of customers. Since 1990, governments in developing and emerging countries have signed more than 260 PPP contracts in the sector, and it is estimated that by 2007, PPP projects were supplying water to more than 160 million people in these countries. Nonetheless, the market share of water PPP projects in developing and emerging countries stood at only about 7 percent of the total urban population, up from less than 1 percent in 1997 and about 4 percent in 2002.

Public-private partnership in urban water supply has been controversial, particularly in recent years, after a series of highly publicized contract terminations raised doubts about the suitability of the approach for developing countries. Differing perspectives have generated a large body of literature with ambiguous and, in some cases, contradictory findings. The divergences stem from several factors, including (a) variations in methodology (such as a detailed case study versus cross-sectional econometric analysis); (b) variations in data availability and reliability; and (c) variations in evaluation frames (many focus on a single topic or set of topics—for example, regulation or tariff setting, poverty targeting, or connection costs—whereas others incorporate more variables but cover a time span of only one or two years).

Some observers have viewed contracting of the provision of such essential services as inherently fraught with conflicts, given the monopolistic nature of these services. Others are more pragmatic but question whether PPP arrangements can work well in the diverse settings of developing countries, pointing to weak institutional capacities, poor governance, and gaps in the rule of law and enforcement of contracts. Others point to a few highly publicized failures as evidence that PPPs as such are not suited to the WSS sector and to conditions in developing countries in particular. Still others attribute those failures to vested interests and political manipulation, and point to selected successes for lessons on how to make PPPs work. Underlying most studies are gaps in data coverage and quality, reflecting the fact that performance data disclosed by water utilities are limited and rarely comparable among utilities and over time. The scarcity of published performance information contributes to an impression of secrecy and lack of accountability, whether for public utilities or private operators.

This study provides objective information and analysis on the performance of PPP projects in urban water supply and sanitation in developing countries. It reviews the spread of urban water PPP projects during the past 15 years, and assesses whether and how they have helped to improve services and expand access for the populations concerned. The study uses a structured framework to assess the performance of more than 65 large water PPP projects that have been in place for at least five years (three years in the case of management contracts) and that provide services to a combined population of almost 100 million (see appendix A). By population size, this sample represents close to 80 percent of the water PPP projects that were awarded before 2003 and have been active for at least three years. The analysis focuses on the actual impact of these projects for the concerned populations, that is, the net improvements achieved under these partnerships.[1] To the extent that the available data permit, four dimensions of

1. This approach does not look at the operators' compliance with contractual targets, which is an important but altogether different issue and not the subject of this study.

performance are analyzed: access and coverage expansion, quality of service, operational efficiency, and tariff levels. The limitations and pitfalls inherent in the analysis of each performance dimension are addressed.

The term *public-private partnership* is used in different ways in the literature, so it is important to clarify what this report is about. The PPP projects analyzed in this study are those in which the provision of urban water and sanitation services is delegated by contract to a private operator, which usually takes over the management of an existing utility. The sample set covers divestitures (in which infrastructure assets are sold to private investors); concessions (whereby a private operator becomes responsible for both operation and investment); leases-*affermages* (whereby a newly established private utility operates a publicly owned system and collects revenues that it then shares with the public owner, who remains in charge of investment); management contracts (in which the services are provided by a publicly owned utility that is managed by a private operator); and mixed-ownership companies (in which a private investor takes a minority share in a water company and operates it on behalf of the local authorities, sharing the financial gains with the public partner). This report refers to water PPPs for the sake of simplicity—because, in most cases, sewerage services were a secondary (or absent) activity—and data on sanitation are provided where available.

The study does not cover several other schemes involving the private sector, such as contracts limited to bulk facilities for construction, financing, and operation of water purification and wastewater treatment plants (build, operate, and transfer projects, or BOTs and similar arrangements), or technical assistance and service contracts. It also does not include cases in which a portion of public utilities' shares were sold to private investors without transfer of management control to a private party. Finally, only urban water supply and sanitation PPPs serving at least 25,000 people are reviewed, thereby excluding the many small and often informal operators that provide service to many households living in periurban areas in developing countries.

This study is not intended as a policy manual, nor is it a census of all PPP projects operating in the urban water sector. Neither does it presume to systematically assess the relative performance of public versus private modes of service delivery, although in the few cases where such information is available and relevant, comparisons are made. Finally, it is important to keep in mind that a PPP is merely one instrument among many for improving performance outcomes, and its efficacy depends on the presence of a constellation of other measures: sectoral policies, regulatory oversight mechanisms, financing instruments, subsidies, and related poverty targeting programs. Those measures, although important, are not the focus of this study.

Chapter 2 summarizes the historical development of water PPPs in developing countries, reviewing the current state of the market, the rate of contract cancellations, and the evolution of the industry. Chapter 3 reviews the performance of PPP projects in terms of access, service quality, operational efficiency, and tariffs. Chapter 4 draws conclusions and lessons on how to make public-private partnership a more viable and sustainable option for improving WSS services in the developing world.

## 2.

# EVOLUTION OF WATER PPPs IN DEVELOPING COUNTRIES

Part of the controversy over private water operators in developing countries has deep historical roots. In the 19th and early 20th centuries, urban water systems in many cities of the Americas and Europe (as well as in colonies or dependencies) were financed, built, owned, and operated by private firms. Many of these private waterworks abused their monopolistic position, often by restricting investment and disregarding service quality. Not surprisingly, this led to the nationalization of water utilities almost everywhere. Two decades ago, private waterworks had all but disappeared, except for a small portion of the markets in the United Kingdom and the United States.

In France and Spain, however, an alternative way of involving private companies in water operations, different from outright privatization, had been emerging over more than a century. The concept was that of a partnership with shared responsibilities, in which local governments delegated the management of a water utility to a private operator while retaining the assets as public property. Various contractual forms evolved, with differing levels of responsibility and risk for the private partner, ranging from concessions to management contracts. In France, the most original concept was probably the *affermage* (a newly established private utility operates a publicly owned system and collects revenues that it then shares with the public owner, who remains in charge of investment), whereas in Spain, *empresas mixtas* (mixed-ownership companies) emerged. Options other than concession schemes leave investments in the hands of the public partner, allowing local authorities to keep control over the expansion of the water system—

a key determinant of urban planning—while freeing them from the daily operation of the utility.

In what is now the developing world, the private sector also played a major role in the development of the first urban water systems, often through foreign investors. During the first half of the 20th century, a similar movement occurred with a return to public management and control of water utilities. In Africa, though, several private water companies remained in place well after independence.

### The Water Sector in the 1990s

By the end of the 1980s, water supply systems in most cities of the developing world were facing growing problems of quality, reliability, and coverage. Burgeoning urban populations required massive investments in expansion, which few public utilities had the means to carry out, and water rationing was becoming the norm. Political interference and clientelism had led to excessive staffing and low morale, which translated into inefficiency and poor service quality. Governments, both local and national, had found it politically convenient to let inflation erode tariffs to levels well below costs. Many utilities, dominated by a civil engineering culture, were more interested in building large hydraulic works than in operating them. Because investment resources came more from central government budget transfers than from tariff revenues, customer service was hardly a priority.

A vicious circle had developed: without maintenance, systems deteriorated, delivery became unreliable, and water quality worsened. Ill-served customers neglected to pay their water bills and resisted tariff increases, leaving even fewer resources. Meanwhile, national budgets in developing countries grew tighter, starving the sector of its traditional source of investment funds. The situation had become untenable by the early 1990s, but the existing state monopolies offered no obvious solutions. Even if politicians could agree that tariffs should be raised to improve service quality and access, they had scant trust that their public utilities would use the proceeds efficiently to deliver improvements for the population. The special interest groups that had come to control many public utilities were loath to change the status quo. Often, households that could afford to had developed alternative sources (private wells) or coping measures (rooftop rainwater-collection tanks, and filters) and, consequently, were reluctant to pay more. The poor, who could not afford such investments, were left to suffer. Central governments and donors became increasingly reluctant to provide public money to the urban water sector. Under such circumstances, replacing the public management of water utilities with private management—under contracts that specified service obligations and financial incentives—appeared to be a sensible way to break the status quo.

## Private Financing of Water Infrastructure

Infrastructure utilities in a few countries, such as Chile, New Zealand, and the United Kingdom, had been privatized in the 1980s, and the early results suggested that privatization might provide a solution for poorly performing public utilities. The rationale was that a private operator would operate more efficiently because of its profit motive and the fact that its contract contained clear, consistent objectives and means rather than the multiple and often conflicting goals assigned to state-owned utilities (Harris 2003). The separation of policy and regulation (which would remain the government's responsibilities) from the provision of services (which would become the responsibility of the private operator) would provide accountability through an arm's-length relationship that was largely missing under public provision. The gains from reforming poorly performing utilities were expected to be large enough to allow private operators to directly finance the investments that were needed to improve service quality and expand access for the poor.

Private participation expanded rapidly in the early 1990s in the telecommunications, electricity, and transport sectors, bringing in massive amounts of private investment. The water supply and sanitation (WSS) sector also appeared to be a good candidate, despite the special challenges involved. Competition in the sector was to be limited to contract awards, because water supply and sewerage services constituted a natural monopoly. Underground assets were difficult to assess, introducing much uncertainty in investment plans. Tariffs tended to be very low (indeed, below cost), and the sector as a whole was beset with entrenched social and cultural issues.

The 1989 water privatization in England and Wales played a major role in convincing policy makers that private financing of urban water utilities could be viable. This privatization was a momentous event for the industry worldwide, raising massive amounts of private money from international financial markets, and a new regulatory scheme was put in place (box 2.1). Many observers expected that such a promising new approach could be replicated on a large scale in the developing world, where a considerable amount of money was needed to fund the investment backlog.

**BOX 2.1**

**Putting in Place a Modern Regulatory Framework: The 1989 Water Privatization in England and Wales**

In England and Wales, 29 private water companies were still in place in the late 1980s, serving about a quarter of the population. The rest of the population received water supply and sewerage services from 10 regional public water authorities, which had been created by the consolidation (around hydrological boundaries) of 1,400 public water and sewerage services through the 1973 Water Act.

In 1989, the regional water authorities in England and Wales were converted into 10 companies operating under private law. The companies were responsible for both water supply and sewerage services in their appointed areas; their shares were sold in a massive public floatation. This was the largest-ever transfer of water and sewerage assets to the private sector, and it adopted a fundamentally different model from that of France and Spain. In addition to granting licenses to operate the systems and collect tariffs from customers, the reform transferred ownership of the water infrastructure entirely to private investors.

The privatization was accompanied by the creation of the Office of Water Services, a special regulatory agency for the sector. The agency was responsible for implementing an innovative regulatory mechanism based on the so-called price-cap methodology, with tariff reviews occurring every five years. For each private company, the regulator set the tariff evolution for the next five years according to efficiency-improvement objectives. If a company was able to cut costs below the level set by the regulator, it could keep the savings it made during that five-year period. Then, at the next five-year tariff review, the tariff would be adjusted so that actual savings could be passed on to customers. This was a major departure from the cost-plus regulation that had been used previously for the economic regulation of private providers.

## Evolution of the Water PPP Market since 1990

The development of water public-private partnerships (PPPs) in developing and transition countries occurred in several phases. The first contracts awarded in the early 1990s generated considerable momentum during that decade. However, as problems started to appear in several high-visibility projects, enthusiasm faded somewhat, and since 2001, the market has gradually taken a different course.

### *The First Wave of Water PPP Contracts*

By the end of the 1980s, large private water operators had mostly vanished from developing countries. The major exception was in Côte d'Ivoire, where the private operator Société d'Amenagement Urbain et Rural (SAUR) had been present since 1960 and operated the national private water utility Société de Distribution d'Eau de Côte d'Ivoire (SODECI) under an affermage contract. Then, in the late 1980s, the government of Guinea requested assistance from the World Bank to replicate the Côte d'Ivoire approach, leading to the award in 1989 of a 10-year affermage contract to SAUR. Under this arrangement, the private operator was to be responsible for improving service quality and efficiency, and the government would remain in charge of investment.

It was in Latin America, however, that the development of water PPPs would get the most impetus. Several large countries in Latin America had undertaken water sector reforms in the 1980s, dismantling their national water utilities to create decentralized bodies at either the provincial level (as in Argentina and República Bolivariana de Venezuela) or the municipal level (Colombia). These reforms had proved largely disappointing, because the transfer to ill-prepared local governments often caused massive reductions in investment. The region was also undergoing widespread economic liberalization at the time, and private sector participation was being introduced on a large scale for infrastructure services. For governments facing considerable investment needs, accessing private capital became the major motive for seeking private sector participation in urban water utilities.

The first water PPP of the period in Latin America was a concession awarded in 1991, for the Argentine provincial utility of Corrientes, to a consortium led by a newly privatized British operator (Thames Water). This rather small venture was followed by two much more ambitious attempts, with tenders launched for the concession of the water utilities of two major capitals: Caracas (República Bolivariana de Venezuela) and Buenos Aires (Argentina). Although the Caracas tender did not go through, the Greater Buenos Aires concession was successfully awarded. The winning consortium took over in May 1993, committing to invest US$4 billion over the 30 years of the contract—an unprecedented amount for the water industry in a developing country.

### *Initial Momentum for Water PPPs*

The award of the Buenos Aires concession generated considerable momentum. The concessionaire achieved early successes after assuming control. The city's recurrent water-rationing problems during the summer months were solved in the first year, and more than a million people were connected to the water network during the first four years of private operation, closing the gap with the national coverage average (Ducci 2007).

A series of PPP contracts for water utilities followed in the next three years, all over the developing world. The most significant included Cancun (Mexico) and Gdansk (Poland) in 1994; Kelantan state (Malaysia) and Santa Fe province (Argentina) in 1995; Senegal, Manila (the Philippines), Cartagena (Colombia), and Aguascalientes (Mexico) in 1996; and Gabon, Cordoba (Argentina), La Paz–El Alto (Bolivia), Budapest (Hungary), Barranquilla (Colombia), and Casablanca (Morocco) in 1997. All except the Cancun concession were awarded to foreign operators. As shown in figure 2.1, the number of new contract awards went up steadily, and the population served by private operators rose from 6 million to 93 million between 1991 and 2000.

More and more governments were embarking on the contracting of private water operators for their urban water utilities. The number of developing countries with water PPP projects increased ten-fold, rising from 4 to 38 between 1991 and 2000.[2] There were initially only a few isolated cases of early contract termination (Tucuman province in Argentina and Kelantan state in Malaysia). During that period, Latin America played a leading role, accounting for 43 million (45 percent) of the 93 million people who were served by private water operators by 2000 in developing countries. By that year, Argentina had become the largest market by far, with as many as 18 million people (more than half the urban population) served by private water operators. Other regions were far behind, with 14 million in Asia (Manila and Jakarta), 16 million in Sub-Saharan Africa, 13 million in Eastern Europe and Central Asia, and just 7 million in the Middle East and North Africa.

### *Expectations for Private Finance Proved Unrealistic*

Publicly reported figures on the private financing of water infrastructure were initially encouraging. Between 1990 and 2000, projects with private investment committed almost US$43.2 billion to water utilities in developing countries.[3] Observers concluded that international financial institutions

2. These are countries with at least one water PPP serving more than 100,000 people. Before 1991, water PPPs (according to the definition of this study) in developing countries were mostly limited to Côte d'Ivoire, Macao (then under Portuguese administration), Guinea (since 1988), and Chile (Lo Castillo, an affluent neighborhood of Santiago).

3. The investment figures recorded in the World Bank/PPIAF Private Participation in Infrastructure (PPI) Projects Database (2007 U.S. dollars) refer to contractual arrangements in which private parties have at least a 25 percent participation in the project contract according to publicly available information. The divestitures covered are those with at least 5 percent of equity owned by private parties. Contracts in which operational control has been transferred to the private sector but without obligation to invest (or with obligation smaller than the threshold already indicated) are not included in the PPI database investment figures.

**Figure 2.1 Water Utility PPPs Awarded and Urban Populations Served in Developing Countries, by Region, 1991–2000**

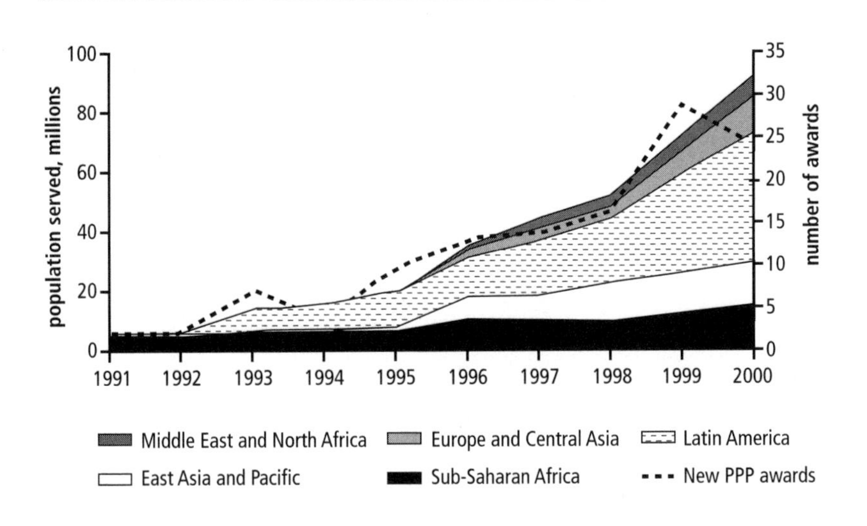


*Source:* Author's calculations based on the Private Participation in Infrastructure Projects Database (PPI database), World Bank/PPIAF.

could reduce sovereign lending to the sector and, instead, provide support through private financing instruments.

These expectations did not materialize. When compared with other infrastructure sectors, private financing of urban water utilities was limited, representing only 5.4 percent of the total investment commitments in private infrastructure between 1990 and 2000. Only a few private companies were participating, and the five most active among them (Suez, SAUR, Veolia, Thames Water, and Agbar) represented 90 percent of the total investment commitment during the period 1991–97. Furthermore, figures on investment commitments were for the total amounts to be invested over the duration of the contracts (often 30 years), and most of the commitments were for a few large projects, with those in Chile, Buenos Aires (Argentina), and Manila (the Philippines) representing nearly half the total amount. Finally, many concessionaires proved unable to borrow on a project finance (nonrecourse) basis from private financiers as originally expected and became constrained by their balance sheet. All in all, much less private investment occurred than was initially expected.

### *Trends since 2001*

The year 2001 was a turning point for water PPPs (figure 2.2), with the fallout from the acute economic crisis in Argentina, which was the largest

Figure 2.2 Water Utility PPPs Awarded and Urban Populations Served in Developing Countries, by Region, 1991–2007

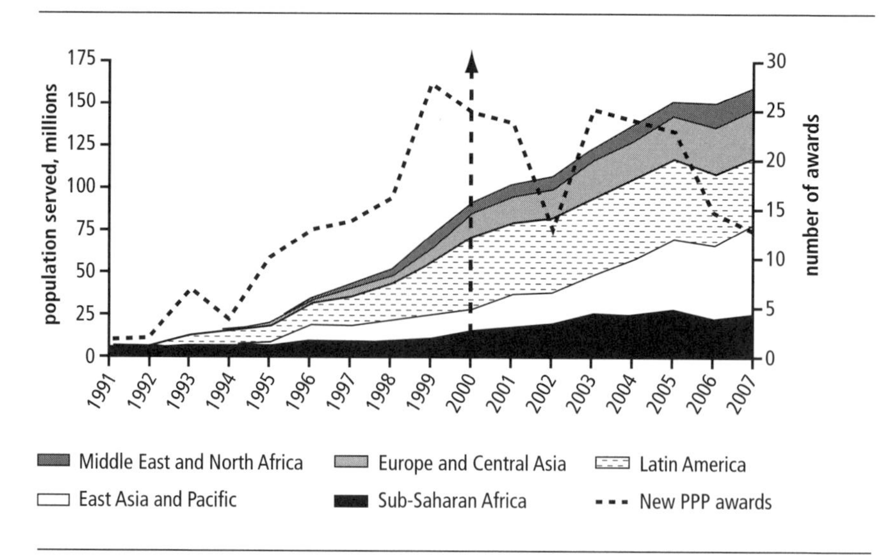


*Source:* Author's calculations based on PPI database.

market for private water operators at the time. The number of contract awards dropped the following year, and in 2003–05, new activity became concentrated essentially in four countries (Chile, China, Colombia, and the Russian Federation). Since 2006, the number of contracts awarded annually has dropped sharply to pre-1999 levels, and the new awards are concentrated in a few countries, with China taking the largest share.

In contrast with the decline in the number of contract awards, the size of the urban population served by private water operators has continued growing. It rose from about 94 million in 2000 to an estimated 160 million in 2007. Two factors lie behind this growth: first, many new contracts awarded in the past five years have been for large urban utilities, and second, many operators have significantly expanded their customer base on existing contracts.

By the end of 2007, there were more than 220 active water PPPs in 41 developing and emerging countries. Water PPP projects have been developing in different ways, depending on the country or region, responding to the specific features of reforms, country risks, and financial markets, and to the local political economy. Between 2000 and 2007, the number of PPP customers fell from 44 million to 39 million in Latin America, but it rose sharply, from 14 million to 50 million, in East Asia, which has now become the biggest market for private water operators. It also rose in all other regions: from 15 million to 25 million in Sub-Saharan Africa, from 15 million to 29 million in Eastern Europe and Central Asia, and from 7 million to 13 million in the Middle East and North Africa.

This evolution was underlined by a gradual change in the financial design of the new contracts awarded. Water PPP projects have increasingly been based on a combination of public and private financing: either in lease-affermage arrangements (Africa and Eastern Europe) or in concessions in which the government provides a sizable part of the capital expenditure financing (for example, Colombia). In contrast with the dominance of the concession model in the 1990s, this second generation of PPP projects relies more and more on public financing for investment. Private financing in water infrastructure is occurring, as seen in PPP projects in Brazil, Chile, China, Colombia, Malaysia, Morocco, and the Philippines, but it is concentrated in countries where operators have gained access to long-term debt in local currency.

**Figure 2.3 Status of Water Utility PPP Projects—Active, Expired, and Terminated, by Region, 2007**

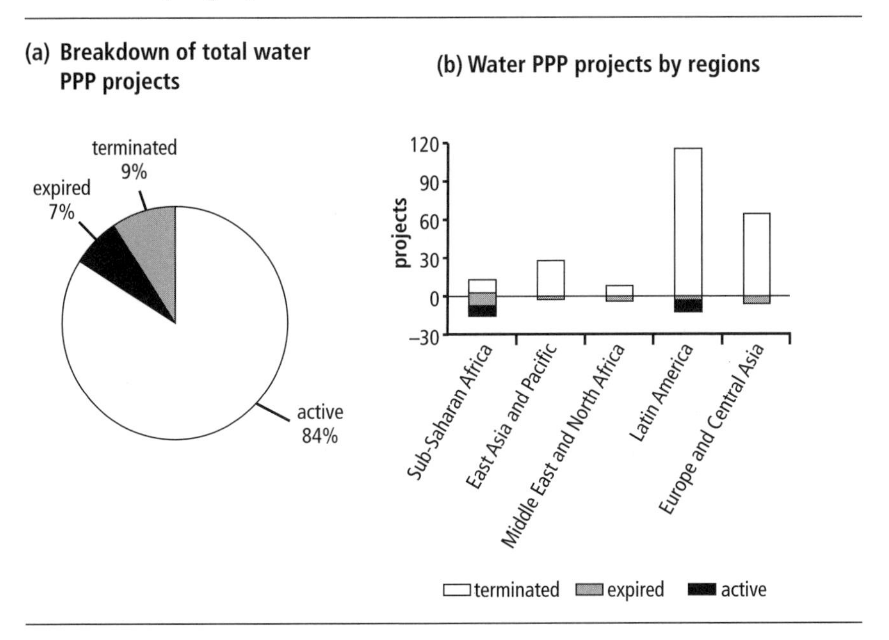


*Source:* Author's calculations based on PPI database.

## Early Termination and Expiration of PPP Projects

The fact that several highly visible water PPPs ran into difficulties that led to the early termination of their contracts has encouraged the perception that many water PPPs in developing countries are encountering problems and being canceled. Indeed, PPPs did not always work as expected in the contractual arrangements, and several projects have failed. But a closer look at the whole picture shows that only a minority of PPPs have been prematurely terminated. Figure 2.3 shows the status at the end of 2007 for all water utility PPPs that have been in place in developing countries since 1990: 228 were active, 18 had expired (and the utility had returned to public management at the end of the contract), and 22 had been terminated early.

### *Most Water PPP Projects Still in Place*

Some 84 percent of the contracts that have been in place since 1990 were still in operation by the end of 2007. Only two (the Mendoza and Catamarca concessions in Argentina) were reported by the PPI Projects Database as being in distress at the end of 2007.[4] Other contracts that had previously been in crisis had been terminated in the previous three years, or the conflicts had been resolved (usually by the exit of a foreign operator and transfer to local investors).

Only 9 percent of the contracts awarded since 1991 have been terminated early following conflicts between the government and the operator.[5] This is a reasonable figure, considering the challenging situations that private operators have often had to face and the importance of the human factor (that is, whether the partners actually get along) in the outcome of any arrangement based on a partnership. Another 7 percent were cases—mostly short-term management contracts—in which the utility returned to public management when the contract expired.[6]

About 205 million people have been served by water PPPs at one point in time during the past 15 years. When the 160 million people who were still being served at the end of 2007 are deducted, about 45 million people (about 25 million for terminated contracts and 20 million for expired contracts) remain who are being served by utilities that returned to public management after experimenting with private operators.[7]

---

4. The PPI database considers a project to be in distress when the exit of the private sector has been formally requested or a major dispute is ongoing.
5. For contracts awarded before 1998, the rate of early termination is 14 percent; for contracts awarded between 1998 and 2002, the rate is 9 percent.
6. PPP contracts that expired but were followed by another PPP arrangement (as when a management contract was replaced by a lease or concession) are not counted in this percentage.
7. This includes several prominent contracts, including the one in Buenos Aires (8.5 million people), which was the largest concession in the developing world and was terminated in 2006.

**Table 2.1 Large Water Utility PPPs That Returned to Public Management between 1990 and 2007**

| Region | Contracts terminated early | Contracts expired without renewal | Population served (millions) |
|---|---|---|---|
| Sub-Saharan Africa | Central African Republic, Chad, Comoros, The Gambia, Mali, Rwanda, Dar es Salaam (Tanzania) | Guinea, Guinea-Bissau, Madagascar, Zambia, Johannesburg (South Africa), Kampala (Uganda) | 17.0 |
| East Asia and Pacific, South Asia | Kelantan (Malaysia) | | 0.5 |
| Middle East and North Africa | Hebron (West Bank and Gaza) | Amman (Jordan), Gaza (West Bank and Gaza), Tripoli (Lebanon) | 3.5 |
| Latin America | Buenos Aires, Santa Fe, Buenos Aires province (2), Tucuman (Argentina), La Paz–El Alto, Cochabamba (Bolivia), Punta del Este (Uruguay) | Guyana, Trinidad, Lara & Monagas (Venezuela, R. B. de) | 20.0 |
| Europe and Central Asia | Antalya (Turkey), Borsodviz (Hungary), Vladivostok, Volgograd (Russian Federation) | Kosovo, Elbasan (Albania) | 4.0 |

*Source:* Author.

*Note:* Only PPPs serving more than 150,000 people are recorded. The concession in West Manila (the Philippines), which was terminated in 2005 but was re-awarded to another private operator the following year, is also not figured.

### *Cancelled PPPs Largely Concentrated in Latin America and Sub-Saharan Africa*

Cancelled PPPs are concentrated in Latin America and Sub-Saharan Africa. The most prominent of the contract terminations and nonrenewals—those of PPPs serving a population of at least 150,000—are listed in table 2.1. In Latin America, half of the cancellations took place in Argentina, whereas the overall rate of cancellation for the region stands at 10 percent, similar to the global average. In Sub-Saharan Africa, approximately half of the PPPs awarded either have been terminated early or have expired with a return to public management—a high rate that can be linked to a challenging environment for reforms. It is also noteworthy that most of the cancelled PPPs in Africa were for combined power and water utilities, in which water was a secondary activity.[8] The rate of active projects for combined water and power utilities is only about 20 percent, with half of the contracts having been terminated early—contrasting sharply with the rate for water-only PPPs, for which about 90 percent of the contracts are still active.

### *Reasons for Early Termination of Contracts*

Most cases of early termination of contracts involved significant noncompliance with contractual obligations by one or both sides, followed by a degradation of the partners' relationship to the point that ending the partnership was the chosen solution. A significant proportion of these PPPs were terminated after having been in place for many years; their termination usually reflected difficulties in adapting the contract over time to changing conditions. Several governments grew dissatisfied with the modus operandi of PPP, and felt they could better solve the problems of the sector by regaining direct public control.

Part of the reason for the large number of terminations of concessions in Latin America probably stems from the rather bullish nature of the market in the 1990s, when a few private operators made overoptimistic offers in order to gain contracts. Several cases of early termination can be traced to contracts whose design was not viable or whose bidding process led to unrealistic financial conditions, or both. An example is Cochabamba (Bolivia), where the contract was awarded following a tender from which all but one company had withdrawn. Substantial tariff hikes were needed to make viable the large investment required from the private operator, something that proved socially unsustainable and brought about the rapid demise of the contract.

PPP projects that expired at the end of their contract duration, and were followed by a return to public management, represent 7 percent of the cases. They must be clearly distinguished from PPP contracts that were terminated early. A large majority of the expired water PPPs were for short-term management contracts, with limited transfer of responsibilities to private operators. The reasons these utilities returned to public management are varied, and not necessarily linked to a failure to improve or to satisfy the government's expectations—as illustrated by the successful management contract in Johannesburg, South Africa (Marin, Mas, and Palmer, 2009).

8. This was the case in Chad, Comoros, The Gambia, Guinea-Bissau, Madagascar, Mali, Rwanda, and São Tomé and Principe (combined population of about 6.5 million).

**Figure 2.4 Urban Populations Served by Private Water Supply Operators in Developing Countries, by Country of Origin, 1991–2007**

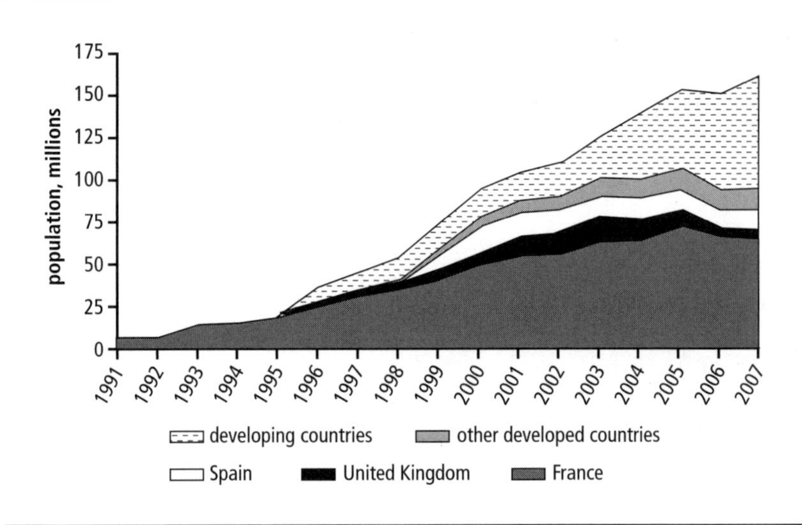


*Source:* Author's calculations.

## New Operators

During the 1990s, tenders for water PPPs typically included rather restrictive prequalification criteria, which often prevented the participation of investors with no previous experience in operating large urban water and sanitation systems. The rationale was that the provision of drinking water was too essential a service to be delegated to inexperienced private investors. Because only a few developed countries had experienced private water operators at the time, the consequence of this cautious approach was that by 2001, five international companies accounted for 80 percent of the population that was served by private operators in the developing world.[9]

The period 2001–06 saw a major change, with the growing participation of new private operators from emerging and developing countries (figure 2.4). Since 2002, the population served by private operators from developing countries has been increasing steadily, accounting for most of the growth observed in this period (see box 2.2). The growth in their customer base represents an additional 55 million people served, according to estimates from this study; in the meantime, the population served by large international operators has remained flat at about 95 million people since 2001.

9. The companies included Suez (36 percent) getting the lion's share, followed by SAUR (15 percent), Veolia (12 percent), Agbar (11 percent), and Thames Water (6 percent).

Box 2.2

**New Private Water Operators from Developing Countries**

Most of the contracts awarded in the early 1990s went to large international operators. The award in 1996 of the Manila concessions required that national investors own 60 percent of the shares of the concessionaire companies, prompting international operators to team up with local partners who had a majority control. In the Manila Water concession (Eastern zone), the partnerships worked out well; the foreign partner transferred know-how, and this led to the gradual emergence of a competent Filipino private water operator, which by 2007 was looking for expansion through PPPs in Asia.

Meanwhile, in Latin America, national investors initially developed by taking on projects that did not interest the large foreign multinationals. In Argentina, Thames Water left control of the Corrientes concession in 1995 to its local partner. In the following years, Argentine investors won several tenders for provincial concessions (in Salta, Santiago del Estero, Formosa, and La Rioja). A similar phenomenon started in Colombia and Brazil in 1998–99; construction companies that were active in the water sector were awarded PPP contracts following tenders in which only local investors participated. In all these cases, governments chose to ease prequalification criteria to increase competition, resorting to various mechanisms to ensure the winning bidder would be able to operate the water utility. In Colombia, the winning bidders contracted experienced technical staff (often former managers and engineers from public utilities). In Argentina, the investors who won the Salta concession signed a technical assistance contract with an established water utility (SANEPAR, the state-owned water utility of Paraná, Brazil).

National operators gained ground in Latin America after 2000, because the initial experiences with them proved encouraging. Between 2001 and 2004, Colombian investors won almost all the PPP contracts awarded in that country under the Programa de Modernización de Empresas (PME). The same trend happened in Chile during the second wave of PPPs in 2002–04. Several operators became significant national players by winning multiple contracts. In recent years, national investors kept increasing their market share as several international operators were exiting the market. This was the case with Suez in Manaus and Limeira (Brazil) and in Cordoba (Argentina), with Agbar in Campo Grande (Brazil), and with Anglian and Thames Water in Chile.

*(continued)*

Box 2.2

**New Private Water Operators from Developing Countries** ***(continued)***

In other developing countries, different patterns have emerged. A few large private groups directly negotiated PPP contracts for large water utilities. This occurred in Malaysia, where control over the concessions for Johor state (2001), and for Selangor state and Kuala Lumpur (2004) was sold to national companies. In Russia, the development of PPPs since 2003 took place through direct negotiation, involving essentially two companies (RKS and Rosvodokanal) that had links to major energy conglomerates. The Tata group in India is yet another case; it had been providing water utility services in the industrial city of Jamshedpur since the 19th century. The water service had traditionally been provided as a special department of Tata Steel but was established as a separate water company in 2004.

### *Strong Growth of Private Operators from Developing Countries*

By 2007, private water operators from developing countries were serving as many as 67 million people, or more than 40 percent of the market. This figure is an underestimate, because it excludes China, where recorded PPPs serving more than 24 million people are based on mixed control between the international operator and local investors (the latter holding a majority share) and where national operators in small cities may have gone unreported. Neither does it include the two large private operators in Côte d'Ivoire (SODECI) and Senegal (Sénégalaise des Eaux, or SDE), which together serve more than 13 million people. Although controlled by a foreign operator, SODECI and SDE are essentially African companies run by national management and with a large local shareholding.

Many private water operators from developing and emerging countries have now become significant players, and some are now taking a regional view of the market. For instance, Latinaguas (Argentina) won a concession for the city of Tumbes (Peru) in 2005, and ONEP (Office National de l'Eau Potable, in Morocco) won an affermage contract two years later for the national water utility of Cameroon. Malaysian companies have been actively looking for opportunities abroad, even purchasing an English water utility in 2002. Table 2.2 summarizes data from the largest players identified in this study, which serve at least 400,000 people.

### *Partial Retreat of Large International Operators*

The picture is very different when looking at the large operators from developed countries. Overall, the total population served by international operators rose only marginally between 2001 and 2007, from 86 million to

**Table 2.2 Largest Private Water Operators Owned by Investors from Developing Countries (excluding China)**

| Country | Operator (national) | Start year | Main PPP contracts (cities, states, and provinces) | Population served (millions) |
|---|---|---|---|---|
| Malaysia | Puncak Niaga | 2004 | Kuala Lumpur and Selangor state | 6.5 |
| | Ranhill | 2001 | Johor state | 2.9 |
| | YTL | 2002 | Wessex Water (UK) | 2.5 |
| | Salcon | 2005 | Changle (China), Linyi (China) | 0.7 |
| Philippines | Manila Water | 1996 | Manila Eastern zone | 3.0 |
| | DMCI–MPIC | 2006 | Manila Western zone | 3.0 |
| Indonesia | PTJ | 2006 | Jakarta Eastern zone | 3.1 |
| India | Tata group | 2003 | Jamshedpur | 0.4 |
| Russian Federation[a] | Rosvodokanal | 2003 | Orenburg, Krasnodar, Tyumen, Kaluga, Barnaul | 3.0 |
| | EWP | 2005 | Omsk, Rostov | 2.2 |
| | RKS | 2003 | Kirov, Perm, Tambov | 2.0 |
| Morocco | ONEP[b] | 2007 | Cameroon (national utility) | 3.0 |
| South Africa | WSSA-Uzinzo | 1992–2005 | Queenstown, Maluti | 0.6 |
| Argentina | Latinaguas | 1996–98 | Salta, Corrientes, La Rioja, Tumbes (Peru) | 2.0 |
| | Roggio | 2006 | Cordoba | 1.3 |
| | Sagua | 1997 | Santiago del Estero | 0.4 |

| | | | | |
|---|---|---|---|---|
| Brazil | Vega | 2006 | Manaus | 1.4 |
| | Aguas do Brasil | 1999 | Campos, Niteroi, Petropolis (RJ) | 1.3 |
| | Saneatins | 1999 | Tocantins state | 0.9 |
| | Bertin | 2005 | Campo Grande | 0.8 |
| | Odebrecht | 2006–07 | Limeira, Rio Claro (sewerage) | 0.4 |
| Chile | Fernandez Hurtado | 2003 | Esval (Valparaiso), ESSCO | 1.9 |
| | Solari | 2004 | ESSAR, ESSAT, ESSMAG | 1.2 |
| | Luksic | 2003 | ESSAN (Antiofagasta) | 0.5 |
| Colombia | Triple A[c] | 1997 | Barranquilla, Santa Marta, Soledad and others (Atlántico) | 2.2 |
| | EIS | 2006 | Cucuta | 0.7 |
| | Conhydra | 1998 | Buenaventura, Marinilla and others (Antioquia) | 0.6 |
| | Sala | 2003 | Sincelejo, Corozal | 0.4 |

*Source:* Author.

a. For Russian water operators, this table reports the population served under long-term PPP contracts only (several cities have one-year renewable operation and maintenance contracts with private operators representing several more million people served).

b. As a publicly owned national water utility, ONEP also serves about a third of the urban population in Morocco.

c. Triple A has a strategic partnership with the publicly owned water utility of Madrid (Canal Isabel II). It is, in practice, managed by nationals and has been branding itself as a Colombian private company.

95 million people. By the end of 2007, some of these operators who were the most active during the 1990s had significantly retreated from the market (figure 2.5).

Suez, in particular, has left Latin America and refocused on specific markets such as China, Morocco, and Eastern Europe. The one-year drop in the population served by international operators, which took place in 2006, is largely attributable to the disengagement of Suez from Latin America; the termination of its Buenos Aires and and Santa Fe (Argentina), and La Paz–El Alto (Bolivia) concessions in 2006 accounted for a net reduction of about 11.5 million people served. The same year, two of Suez's large management contracts expired—one in Amman (Jordan) and the other in Johannesburg (South Africa)—together accounting for an additional decrease of 5 million people served. Suez also transferred control of the concessions of Cordoba (Argentina) and Manaus and Limeira (Brazil) to its local partners. By 2007, Suez's presence in developing countries was much reduced, after many years of dominating this market. Its only large remaining operations in developing countries outside of China were in Casablanca (Morocco), Algiers (Algeria), Cancun (Mexico), and West Jakarta (Indonesia).[10] English water companies also significantly reduced their presence in developing countries.[11]

In contrast, other international companies have been quite active, not just in the booming Chinese market but also in low-income countries. Over the past five years Veolia has significantly expanded its presence in emerging and developing countries. Its total population served grew threefold between 2000 and 2007, from 7 million to 30 million. Notably, it started to expand its operations in emerging countries mostly after the so-called market downturn of 2001.[12] Several water operators from Western Europe have also been awarded PPP contracts: namely Acea (Italy), Gelsenwasser and Berlinwasser

10. Though Suez had always been the main, albeit minority, shareholder in Agbar, it acquired control in 2007 by taking a majority shareholding. After consolidating Agbar, the market share of Suez in developing countries became comparable to that of Veolia. By the end of 2007, Agbar's main contracts abroad were in Cartagena (Colombia), La Havana (Cuba), Saltillo (Mexico), Greater Santiago (Chile), and Oran (Algeria).

11. The largest U.K. operator, Thames Water, announced that it would withdraw from developing countries, followed by a decision by its owner, RWE, to refocus on energy. Thames transferred the Eastern Jakarta (Indonesia) concession to local investors in 2006 and sold its Chilean water utilities (serving about 2.6 million people). Anglian Water sold its shares in water utilities in the Czech Republic and Chile (ESVAL Valparaiso) in 2003–04.

12. Veolia achieved this growth not only by concentrating on high-end markets (such as Eastern Europe and China), but also by being active in poor countries. It won an affermage contract in 2001 for the national water utility in Niger (one of the poorest countries in the world), and another affermage in 2005 for the city of Yerevan (Armenia). It had submitted unsuccessful bids for PPPs in Madagascar and Ghana (2005) and Cameroon (2007), and won a small concession in Latin America through its partnership with the Spanish group FCC (Fomento de Construcciones y Contratas) in 2005 (San Andres, Colombia).

**Figure 2.5 International Operators in Water Utility PPPs in Developing Countries, 1991–2007**

**(a) Urban population served by main private operators**

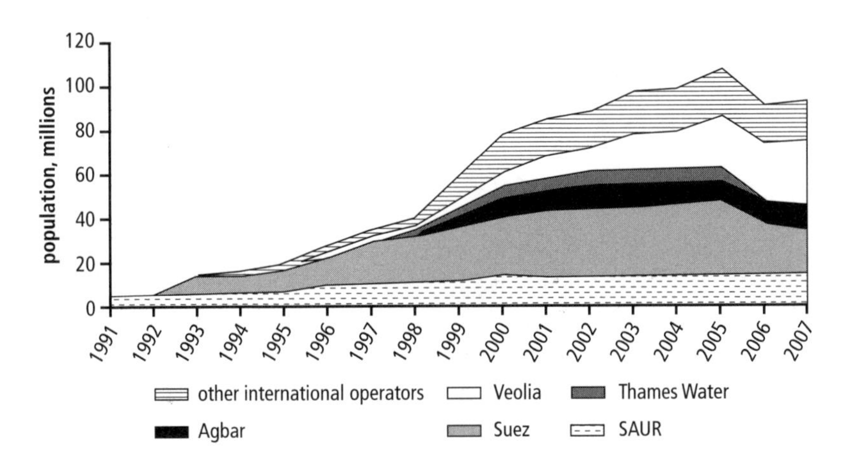


**(b) Repartition of populations served by main private operators, 2007**

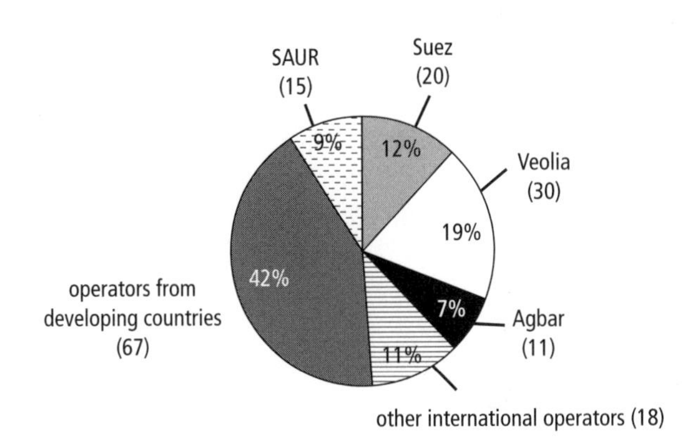


*Source:* PPI database and author's calculations.

*Note:* Population in millions is indicated in parenthesis; percentages are rounded.

(Germany), Aguas de Portugal (Portugal), Stockholm Water (Sweden), and Vitens (the Netherlands). Many of these operators are publicly owned utilities that were not interested in the concession model of the 1990s but are now entering the market through management contracts.

## Conclusions from Trend Analysis

Even though the number of new contract awards has been going down in recent years, the paradox is that more and more people are being served by private water operators, and more and more countries have been introducing water PPPs. During the five years from 2002 to 2007, private water operators have made significant inroads in big countries such as Algeria, China, Malaysia, and Russia. Contracts have been awarded for the first time in the Arab peninsula, Cameroon, Georgia, Ghana, and Peru. Even the failed Manila West concession in the Philippines was rebid successfully at the end of 2006, and the series of contract cancellations in Latin America now seems to be over. The development of water PPPs has not been uniform throughout the developing world; although the PPP approach has suffered setbacks in Latin America, other regions have been gradually adopting it.

In many countries, water PPPs seem to have withstood the test of time. By the end of 2007, 44 developing and emerging countries had active urban water PPP projects. In Armenia, Cameroon, Chile, Côte d'Ivoire, the Czech Republic, Gabon, Ghana, Malaysia, Niger, and Senegal, the majority of the urban population is now served by private operators. In several other countries, private operators serve close to or more than a third of the urban population; those countries include Algeria, Colombia, Cuba, Ecuador, Hungary, Morocco, and Mozambique. Even Argentina still has more than 10 water concessions serving 20 percent of the urban population.

However, about one-third of the developing countries and economies that had water PPP projects during the past 15 years decided to revert to public management.[13] This is a significant proportion, which underlines the fact that PPPs are complex and risky endeavors.

Since the late 1990s, governments and other stakeholders in urban water PPP projects have gradually learned what works and what does not and reflected these lessons in a move away from pure concessions and toward partnerships that rely more on public funding. At the same time, new private operators have entered the market. Many of the newcomers are from developing countries, and they are radically changing the face of a market that, during the 1990s, looked like an oligopoly among a few multinationals. A more mature environment is appearing, more attuned to the needs and specificities of the developing world.

13. Most of these projects were in Sub-Saharan Africa (Central African Republic, Chad, Comoros, Guinea, Guinea-Bissau, Madagascar, Mali, Rwanda, São Tomé and Principe, Tanzania, Uganda, and Zambia), including seven for mixed water-electric power utilities; followed by Latin America (Belize, Bolivia, Guyana, Trinidad, Uruguay, and República Bolivariana de Venezuela); Middle East (Jordan, Lebanon, and West Bank and Gaza City); and Eastern Europe and Central Asia (Kosovo and Turkey).

This overall finding—that water PPPs have been spreading constantly in the developing world during the past 15 years despite difficulties—raises a fundamental question: how and to what extent is this development associated overall with good performance by private operators? That is the subject of the next chapter.

# 3.

# PERFORMANCE AND IMPACT OF WATER PPP PROJECTS

For a better understanding of whether public-private partnerships (PPPs) are a viable option to improve the performance of water utilities in developing countries, this chapter examines whether and how water PPPs have contributed to improving service. This is not an easy task: performance is multidimensional, and measurement of each dimension is fraught with pitfalls. In consideration of these factors, this study uses a case study approach to systematically review a large number of PPP projects. The focus is on assessing whether the partnerships brought significant net improvements to the population, compared with the previous situation, in each of four dimensions of performance: (a) access to water and sanitation services, (b) quality of services, (c) operational efficiency, and (d) impact on tariffs.

In practice, the performance of a PPP project depends on the actions of both the private operator and the contracting government, with the government playing a more or less important role depending on the PPP scheme adopted. Different contractual schemes cannot be expected to achieve the same things. This is especially the case with management contracts, which are typically of short duration and entail only a limited transfer of control to the private operator. The analysis of projects distinguishes management contracts from long-term PPPs (divestitures, concessions, leases-*affermages*,[14] and mixed-ownership companies) in which services are

14. A newly established private utility operates a publicly owned system and collects revenues that it then shares with the public owner, who remains in charge of investment.

provided by a newly established utility, partly or fully owned by a private operator.

## Evidence from the Literature

Even though a rather large body of literature has been published on PPPs, the existing evidence seems somewhat confusing. Part of the confusion stems from the many challenges associated with assessing the performance of private operators. This study adopted a methodology that is a middle ground between individual cases studies and econometric studies.

### *The Challenge of Assessing the Performance of Water PPPs*

The effect of public-private partnerships on the extent and quality of water supply and sanitation services has been much debated. Have PPP projects improved the service for existing customers? Have they extended services to those previously lacking access? Have they improved operational efficiency? What is the effect on tariffs? These questions are essential, but they are difficult to answer in practice, given some of the sector's basic characteristics, which include the following:

- *The ambiguity of performance indicators.* The limitations of commonly used indicators are a persistent source of difficulty. These indicators are often based on rough estimates and are calculated differently from one country or utility to another. Measuring water losses is notoriously complex, but even a seemingly straightforward indicator such as service coverage can be hard to estimate accurately (box 3.1).
- *The multidimensional nature of performance.* The various components of performance are linked: one cannot meaningfully discuss an increase in the number of connections while ignoring service interruptions, or discuss a tariff raise without reference to possible improvements in access and service quality. Parameters interact as well: one cannot understand water losses without simultaneously considering service continuity. The presence of sanitation services further complicates the picture, because data that distinguish between the two services are rarely available.
- *Influence of multiple local factors on operating costs and wide variety of tariff structures.* Operating costs of water utilities are largely determined by local factors such as availability of water resources and topography. A wide variety of tariff structures can be found among utilities, with consumption brackets, fixed charges, and billing sometimes based on estimates of consumption. The size of a utility also plays a significant role through economies of scale, because fixed costs represent a large portion of the cost structure. All these factors greatly complicate the task of comparing costs and tariffs across water utilities.

Box 3.1

**Coverage Ratio: The Challenge of Estimating an Apparently Simple Indicator**

The coverage ratio is computed starting from the number of active residential connections, which can be obtained from the utility's customer database. Then one needs the figures for the total population in the service area and the number of people per household—figures for which only broad estimates can be obtained in most developing countries. The average number of people per household varies across countries and, within a given city, across social categories (with poor households tending to have a higher birthrate and a larger extended family under the same roof). The estimate of total population is also usually a broad one: official censuses are often carried out only once a decade, and fast urban growth makes such figures unreliable.

A further issue is what figure to put in the numerator and denominator when calculating coverage. Illegal connections are numerous in some cities, but whether they should be included in the coverage figure is not obvious, and their number is hard to assess. Many countries in Sub-Saharan Africa use a ratio of 8–10 people per individual connection, to account for piped water provided to neighbors through resale. Some utilities report access based only on the population in "authorized" urban areas, leaving out the poor living in illegal slums, which results in overestimated coverage figures.

Another major issue is what is meant by *access*. The Joint Monitoring Program, which tracks progress toward the Millennium Development Goals (MDGs), uses a criterion of "improved access," meaning access to either an individual connection or a standpipe located within 200 meters of the household. In many published studies, it is not always clear what criterion is used—despite the obvious differences in the quality of service provided and investment costs. Estimating the actual number of people served under the improved access criterion is even more difficult than for household connections. In cities where a large portion of the urban population accesses piped water through community standpipes or by purchases from neighbors, the coverage ratio reported is sometimes no more than a guess.

- *The difficulty of obtaining performance data on water services.* Perhaps the most important impediment to meaningful analysis is the widespread difficulty of obtaining good performance data in the sector, whether from public or private providers. Many public water utilities in the developing world lack a proper framework for performance monitoring, and often the data they report are unreliable. As a consequence, many projects have lacked an appropriate baseline against which to measure performance after the transfer to a private operator. Neither is information on PPP contracts always easily accessible to the public.

*Findings from the Literature*

Despite the data limitations, a rather large body of literature has been published on water PPPs. Studies on the impact of private sector participation fall into two groups: case studies of individual projects, and econometric analyses of multiple utilities.

Case studies typically look at how the performance of a water utility evolves after the entry of a private operator. They have tended to concentrate on a few cases (Senegal, Buenos Aires [Argentina], and Manila [the Philippines]), leaving many others (such as Côte d'Ivoire, Casablanca [Morocco], Guayaquil [Ecuador], and Jakarta [Indonesia]) largely undocumented. A desk review of published case studies on water PPPs (Clarke, Kosec, and Wallsten 2004) identified 25 projects in developing countries and concluded that private participation had a broadly positive impact in 16 cases, a negative impact in 5 cases, and mixed results in 4 cases. Especially visible were improvements in coverage, labor productivity, and the quality of services, which were often associated with tariff increases.

Econometric studies use larger samples, often attempting to compare the performance of public and private utilities, but they face the same limitations of sample selection. To reach meaningful conclusions, they need to use a large enough sample to be able to control for the many external factors that can influence performance, but early studies often had to rely on small samples, representing only a small number of countries or utilities over short periods of time. The criteria for classifying utilities as private is not always clear or even relevant. This affected the robustness of their findings. A second group of studies tried to overcome the limitations that were imposed when using data obtained from utilities, by relying instead on data from national household surveys, as with Clarke, Kosec, and Wallsten (2004) on Argentina; and Gomez-Lobo and Melendez (2007) and Barrera and Olivera (2007) on Colombia.

The findings of these studies are summarized in table 3.1. Considering each dimension separately, the findings remain inconclusive. Overall, the average impact of PPPs appears neutral on access and coverage, positive on service quality and operational efficiency, and rather inconclusive on tariff levels. Notably, none of the studies covers all four of the dimensions of utility performance—access and coverage, quality, efficiency, and tariffs.

Two recent studies by the World Bank were able to rely on samples large enough to ensure that exogenous factors could be properly controlled.

**Table 3.1 Summary of Study Findings on the Impact of PPPs on Water Utility Performance**

| Region | Study | Access or coverage | Service quality | Operational efficiency | Tariff level |
|---|---|---|---|---|---|
| Africa | Estache and Kouassi (2002) | n.a. | n.a. | Positive | n.a. |
| Africa | Kirkpatrick, Parker, and Zhang (2004) | n.a. | n.a. | Inconclusive | n.a. |
| Argentina | Galiani, Gertler, and Schargrodsky (2005) | Positive | Positive | n.a. | n.a. |
| Latin America (mostly Argentina) | Clarke, Kosec, and Wallsten (2004) | Neutral | n.a. | n.a. | n.a. |
| Argentina | Maceira, Kremer, and Finucane (2007) | Neutral | n.a. | n.a. | n.a. |
| Asia | Estache and Rossi (2002) | n.a. | n.a. | Neutral | n.a. |
| Bolivia | Barja, McKenzie, and Urquiola (2005) | Positive[a] | n.a. | n.a. | Inconclusive |
| Brazil | Rossi de Oliveira (2008) | Inconclusive | n.a. | Positive | Higher |
| Brazil | Serão da Motta and Moreira (2004) | n.a. | n.a. | Positive | Neutral |
| Chile | Bitran and Valenzuela (2003) | n.a. | n.a. | Positive | Higher |
| Colombia | Gomez-Lobo and Melendez (2007) | Inconclusive | Positive | n.a. | Inconclusive |
| Colombia | Barrera and Olivera (2007) | Inconclusive | Positive | n.a. | Neutral |
| Hungary | Boda and others (2008) | n.a. | n.a. | Unclear | Neutral |
| Malaysia | Lee (2008) | Inconclusive | n.a. | n.a. | Inconclusive |
| World | Ringskog, Hammond, and Locussol (2006) | Greater | Positive | Positive | Inconclusive |

*Source:* Author.

n.a. = not applicable.

a. An earlier publication by Barja and Urquiola (2003) found inconclusive results on the impact on access.

The studies adopted different but complementary approaches and covered all dimensions of performance simultaneously. Andrés, Guasch, and others (2008) focused on Latin America and compared the performance of 49 water utilities in seven countries, before and after the introduction of a private operator. Gassner, Popov, and Pushak (2008a) used a large sample to make a meaningful comparison of the performance of private and public water utilities. They compared the performance by building a data set of 977 water utilities in developing countries, of which 141 had some form of private participation. Only those PPPs with data covering at least three years of private operation were included, and the water utility sample contained 6,079 firm-year observations over the 1992–2004 period. To avoid bias in data reliability, the study focused on comparing private utilities with only the public utilities that had been corporatized or were operating under a comparable framework.

Both studies found improvements in service quality and operational efficiency with private operators. On tariff levels, Andrés, Guasch, and others (2008) found that the introduction of a private operator usually resulted in tariff increases, which is not surprising, because most Latin American utilities had tariffs below cost-recovery levels in the early 1990s. However, Gassner, Popov, and Pushak (2008a) concluded on their larger sample that there was no significant difference in tariff levels between PPPs and comparable public operators over the same period.

The main divergence between the two studies concerns the utilities' performance in expanding access. Andrés, Guasch, and others (2008) found that, in Latin America, introducing a private partner did improve coverage, but the improvement was essentially for the transition period (one year before and two years after the takeover). Working with a larger sample that also represented other regions, Gassner, Popov, and Pushak (2008a) found that private operators had performed better when measured by growth in the number of connections, both during the transition period to private operation and after, but their results regarding residential coverage were inconclusive.

### *Methodology Applied in This Study*

It is worth noting that this study has adopted a different methodology from that found in the literature to date. In view of the respective limitations of both the case study and econometric approaches, a choice was made to seek a middle ground between these two methodologies, so as to draw meaningful conclusions despite the many difficulties related to data collection and interpretation. First, this study reviewed a large number of projects, with a focus on performance indicators, thereby providing a more complete picture than has usually been provided in individual case studies. Second, the performance data and analysis have been linked to well-identified projects, contrary to the "black box" approach typical of econometric studies. The source for each project's performance data is indicated in appendix A.

## Access

The analysis in this section focuses on long-term PPPs, including leases-affermages, mixed-ownership companies, and concessions. Management contracts, which are short-term arrangements typically focused on improving service quality and operational efficiency, are not discussed. The PPPs reviewed in the study represent a combined population of more than 65 million people served. This is equivalent to more than 75 percent of the population served by long-term PPPs that (a) lacked access to water supply and sewerage services at the time the private operator took over, (b) were signed before 2003, and (c) had at least three years of private operation.[15]

### *Experience with PPPs in Latin America*

Latin America provides, by far, the largest sample to analyze the performance of long-term water PPPs to improve access and coverage. This section reviews cases of the various concessions in Argentina, the concession in La Paz–El Alto in Bolivia, the concession in Guayaquil in Ecuador, the various water concessions in Brazil, and the concessions and mixed-ownership schemes in Colombia.

#### Concessions in Argentina

Argentina was the first country in Latin America to embrace private participation in large-scale water operations, and by 2000, it had some 17 million people served by private water operators. In 1990, 76 percent of Argentina's urban dwellers had access to piped water through a household connection, and 39 percent were connected to a sewer network. Only moderate progress of 7 percentage points for both services had been achieved by 2004 (see figure 3.1).[16] The figure also shows the performance in service expansion of five of the seven largest concessions (those active for at least four years and serving more than half a million people).[17] The increase in the national urban average between 1990 and 2004 is provided for reference.

---

15. For reference, the size of the population served will be provided in parenthesis for each project under review. It corresponds to the estimate of the number of people actually served by the water utility, as opposed to those living within the contract area. The figures used are either for 2007 or for the last year the contract was in operation.
16. The data are based on the WHO/UNICEF Joint Monitoring Program for Water Supply and Sanitation. National coverage data mentioned elsewhere in the report come from the same source, unless otherwise noted.
17. The concessions in Tucuman and Buenos Aires (Azurix) provinces, which lasted only about two years and did not achieve any significant expansion in access, are not considered.

Figure 3.1 Increases in Coverage under Five Concessions Compared with the National Increase in Argentina

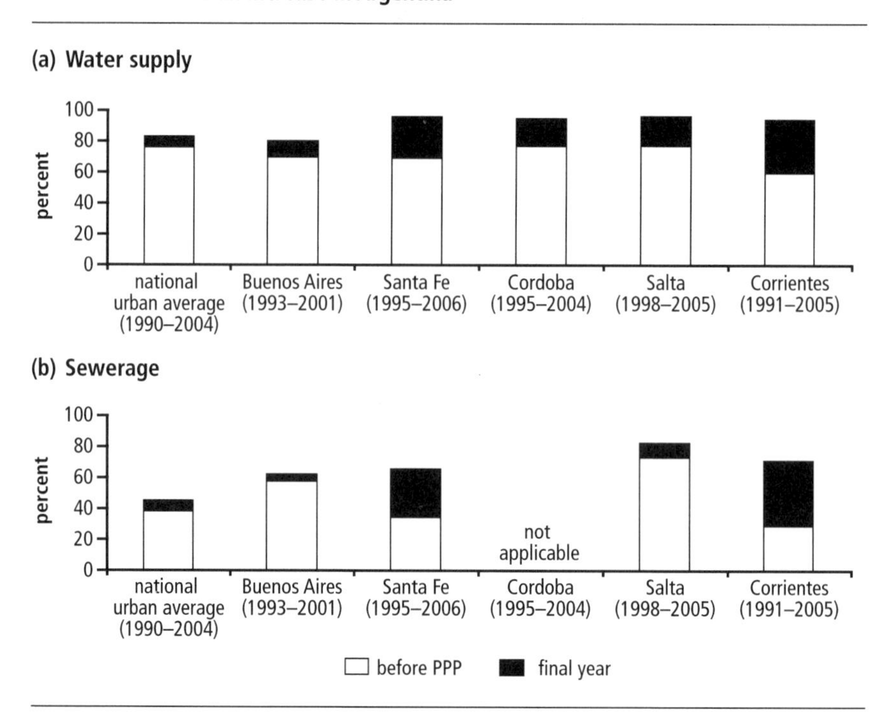


*Source:* WHO/UNICEF Joint Monitoring Program for Water Supply and Sanitation.

*Note:* Coverage at takeover by the concessionaire is compared with the last year of available data, except in the case of Buenos Aires, where the end data are for 2001 (little progress in coverage occurred afterward). Data were not available for Mendoza (active since 1998 but currently in crisis) and the Buenos Aires province concession operated by Aguas de Bilbao (terminated in 2006).

For water supply (figure 3.1a), all five concessions have expanded access significantly. The concessionaires in Santa Fe, Cordoba, Salta, and Corrientes stand out for achieving coverage of about 95 percent, well above the national average, but the progress in Greater Buenos Aires does not appear much different from the evolution of the national average. For access to sewerage services (figure 3.1b), only Santa Fe and Corrientes stand out.

The concession in Greater Buenos Aires (serving 8 million people) served one-third of the country's urban population. During the first five years of the contract, the concessionaire invested substantial amounts in infrastructure expansion and rehabilitation. Between 1993 and 1997, 244,000 water connections were installed, providing access for about 1.5 million people, mostly in poor neighborhoods at the periphery of the metropolitan area. Starting from 10 percentage points below the national average, the gap was closed as coverage went up from 70 percent to 80 percent between 1993 and 1997. Meanwhile, sewer coverage went up from 58 percent to 63 percent, in line with the national average (Ducci 2007).

This initial good performance was not sustained. No further sizable improvement in coverage was achieved after 1998, and it appears that the concessionaire failed to sufficiently invest in network expansion to reach the neighborhoods farther away in the periphery (Delfino, Casarín, and Delfino 2007). This setback is disappointing, because a major contract renegotiation had taken place in 1998 to facilitate service expansion, transferring a large portion of financing the cost of network expansion from new to existing customers and significantly increasing the concessionaire's revenues.[18] Though coverage targets were also revised downward, the concessionaire failed to meet them.

The performance of the provincial concessions in **Santa Fe** (serving 1.9 million people), **Salta** (1.2 million), and **Corrientes** (0.6 million) was more satisfactory. In Corrientes[19] and Salta, coverage continued to expand even after the 2001 economic crisis. The provincial government in Salta contributed part of the investment, and the concessionaire also negotiated financial contributions from municipalities, as well as in-kind contributions from communities, to reduce the impact of expansion tariff. In Santa Fe, contractual targets for expansion and investment were not fully met despite the sizable progress made in extending access to both water and sanitation; the concession was terminated together with the neighboring one in Buenos Aires in 2006 (Ducci 2007).

The concession in the city of **Cordoba** (serving 1.3 million people) was more limited in scope and performed well. Sewerage services remained with the municipality, and the contract made the concessionaire responsible for financing only the primary network expansion, while the municipality and/or users were responsible for financing secondary network expansion. Contractual targets for expansion were fully met and even exceeded. Water

18. The concession was originally designed so that the cost of service expansion would be borne by new customers, who had to pay an infrastructure charge covering the cost of connection plus a portion of the cost of expansion. The amount was considerable, averaging US$415 for a new water connection and US$606 for a new sewer connection (WSP 2001), and proved very unpopular. The new mechanism replaced the infrastructure charge with a bimonthly universal service and improvement charge (SUMA), equivalent to a tariff surcharge and to be paid by all connected customers.

19. Corrientes is the oldest concession active in Argentina, dating from 1991. Its good performance in improving access has been consistent over more than 12 years of private operation. During the first four years, when Thames Water was operating the concession, coverage rose from 70 percent to 85 percent for water, and from 32 percent to 56 percent for sewerage. The trend has been maintained since national investors took over in 1996, reaching 94 percent for water and 72 percent for sewerage in 1995.

coverage continued to increase even after the 2001 crisis, rising from 90 percent in 2001 to 95 percent by 2004 (Ducci 2007).

The concession of Aguas de Bilbao in part of the **Province of Buenos Aires** operated from 2000 to 2006 and proved disappointing. The concession area comprised 13 municipalities with about 1.7 million people and had one of the worst poverty rates in the province.[20] The access figures in 2000 were very low: water coverage was 35 percent and access to sewerage was a mere 13 percent. The concessionaire took over as the financial crisis was unraveling and found itself unable to borrow for investment. The concession was terminated in 2006, with little apparent progress in improving access to water and sewerage.

It is difficult to make a judgment on the overall performance of private concessionaires in Argentina for service expansion. The end of the Argentine peso–U.S. dollar parity in 2001, and the subsequent economic crisis, bankrupted the major concessions whose debt was largely denominated in foreign currencies. Though several concessions have performed well, the largest, in Greater Buenos Aires, failed to make progress after 1998—well before the country's economic crisis erupted. Three econometric studies that focus on Argentina provide different findings: Galiani, Gertler, and Schargrodsky (2005) found that private utilities outperformed public ones, while both Clarke, Kosec, and Wallsten (2004) and Maceira, Kremer, and Finucane (2007) reported no significant difference.

### The La Paz–El Alto Concession in Bolivia

The concession of La Paz–El Alto (serving 1.5 million people), which was terminated in 2006, has been much publicized. Of the populations of La Paz and El Alto, 50 percent and 80 percent, respectively, are considered poor, and the PPP was specifically designed with coverage expansion for the poor as its main contractual objective. Before the concession was awarded, the tariff structure had been revised to significantly lower the monthly water bills of small residential consumers, while raising tariffs substantially for other customer categories. The contract was awarded on the basis of the largest number of new water connections in El Alto. The winning bidder offered to install 72,000 water connections in El Alto in the first five years.

At the beginning of the contract, water supply coverage was about 84 percent in La Paz and 71 percent in El Alto; sewer coverage stood at 66 percent and 30 percent, respectively (Ducci 2007). These rates were for

20. The province tendered for its water services in five distinct zones, using a complex methodology to ensure it would have at least two different operators in separate zones for future benchmarking regulation. The consortium led by Azurix won the concession in the four richer zones, while a consortium led by Aguas de Bilbao, a publicly owned utility from Spain, won the concession in the fifth zone.

the whole area of the two municipalities, as distinct from the concession area, which covered La Paz but only a portion of El Alto. The contract required that within the service area of the concession, access to piped water through household connections be raised to 100 percent in La Paz and 82 percent in El Alto by 2001. In the same year, the target for sewerage was 82 percent in La Paz and 41 percent in El Alto (Ducci 2007).

The level of coverage that the concessionaire achieved has been controversial, especially because of disagreement over whether coverage was calculated for the entire municipal boundaries or just for the service area of the concession. Bolivia's national regulator reported that by 2001, coverage within the concession area was at 99 percent for water and 79 percent for sewerage (Ducci 2007). Overall, the concession brought a sizable improvement in access, with at least 400,000 people—mostly poor families in El Alto—gaining access to piped water.[21]

Despite this good performance, the access issue was the main reason why the concession was terminated in 2006. As many people were gaining access to the network, those living outside the service area of the concession began to protest about being excluded, and a campaign developed to pressure the concessionaire to connect them too. Such claims were reinforced by controversies about whether the concessionaire had fully met its contractual obligations (the number of water connections installed in El Alto, at 53,000, was below the tender proposal) and about the way the regulator calculated coverage (Ducci 2007). The popular discontent was aggravated by the regulator's decision in 2002 to increase connection fees for new customers while leaving tariffs untouched.

The early termination of the La Paz–El Alto concession is a paradox. The PPP performed rather well in expanding access for the poor, in one of Latin America's poorest countries and without recourse to public money for investment. The fact that the partnership ultimately proved unsustainable holds important lessons. First, to make possible the private financing of expansion through international financial markets, the parties had designed the contract with the customer tariff indexed to the U.S. dollar, but this mechanism proved unsustainable. Second, this experience showed the limitations of relying on cross-subsidies to finance access expansion for the poor. The low tariff for small residential customers meant that the operator was serving them at a loss and, therefore, lacked financial incentives to extend services to such customers beyond its strict contractual obligations (that is, in this case, the geographical boundaries of the concession).

21. The figure reported by Ducci (2007) is 600,000 people gaining access, but this is based on rather broad estimates of the number of people per connection, and the 400,000 figure appears to be more realistic.

### A Concession with Public Funding in Guayaquil, Ecuador

The concession in Guayaquil has not been much publicized even though it is, after Buenos Aires, the second largest in Latin America by population served.[22] Guayaquil is Ecuador's economic capital and home to 2.4 million people, or one-third of the national urban population. When the concessionaire took over in 2001, water coverage in the city lagged far behind the national average: only 60 percent of residents had household connections in 2000, compared with the national urban average of 81 percent in 1998. The gap was smaller for sewerage, with coverage of 56 percent compared with a national urban average of 61 percent.

The concession performed remarkably well in expanding access to piped water through household connections. Starting from 245,000 connections in 2000, the concessionaire installed 160,000 new connections in the first five years of operation—equivalent to more than a 10 percent annual increase and three times the contractual target of 55,000. Those gains brought water coverage in the city up to 82 percent in 2005 and benefited about 800,000 people who gained access, most of them living in poor neighborhoods not previously served by the network. This achievement is even more notable, because at the national level, progress in extending urban water access stagnated in the same period. The concessionaire's performance in improving sewerage access was more modest, raising coverage from 56 percent to 62 percent.

The good performance for water access can be largely attributed to a special tax transfer mechanism (the telephone tax), which the central government had introduced back in the 1980s to subsidize new water connections nationwide. It was based on a 10 percent tax on telephone bills, the proceeds of which were granted to utilities to support expansion of the water network in uncovered urban areas. New water connections were provided free of charge to households in urban areas not previously covered by the water network (where most of the population is poor), and part of the cost for the utility of expanding the network was also subsidized. Most new water connections installed by the concessionaire in Guayaquil were financed by this mechanism.[23] Sewer connections were not eligible, explaining why the progress there was much more modest.

---

22. Santiago de Chile (5.3 million customers) is larger but is operated under a divestiture scheme (private ownership of infrastructure).

23. Of the US$85 million of the civil works done by the concessionaire during the first five years, US$39 million was funded from the grants of the telephone tax (Yepes 2007).

**Water Concessions in Brazil**

Between 1995 and 2006, some 36 concessions for municipal water and sewerage services were awarded in Brazil to private operators, serving an urban population of about 6.5 million. Analyzing the evolution of access to piped water in Brazil is made difficult by the fact that utilities typically calculate the coverage ratio on the basis of the population living in "authorized" urban areas—a reference point that excludes illegal slums and that tends to change every year as new neighborhoods are legalized.[24]

The concession in **Manaus** (serving 1.6 million people) is Brazil's largest, and its record in improving access to piped water seems to have been fair. Based on company data, water coverage went up from 72 percent to 86 percent of the population between 2001 and 2005. Ducci (2007) reports that water coverage within the service area was 96 percent in 2006. By then, an estimated 300,000 people had gained access to water. The improvement in sewer coverage was more modest: up from 3 percent in 2001 to just 11 percent in 2006.

The second-largest private operator in Brazil is Aguas do Brasil, a national consortium of construction and engineering firms. According to public data, water coverage in **Niteroi, Campos,** and **Petropolis** (serving 1.1 million people in the state of Rio) rose significantly between 2000 and 2005: from 85 percent to 100 percent in Niteroi, from 75 percent to 96 percent in Campos, and from 70 percent to 80 percent in Petropolis.[25] Meanwhile, the water coverage performance of Companhia Estadual de Aguas e Esgotos (CEDAE), the public utility of the state of Rio, stagnated at about 85 percent.

Data available for 2000–05 on a few medium-size concessions suggest that private operators often performed comparably with public state water utilities. In **Itapemirim** (200,000 people), coverage went up from 86 percent to 98 percent for water and from 72 percent to 90 percent for sewerage, while the water coverage reported by the state public utility Companhia Espírito Santense de Saneamento (CESAN) remained at a high 97 percent. In **Paranagua** (130,000 people), the concessionaire improved coverage from 93 percent to 96 percent for water and from 30 percent to 68 percent for sewerage—a performance comparable to that of

24. It is not uncommon for Brazilian water utilities to report coverage above 100 percent in some years. This reflects a situation where several neighborhoods are being served even though they are still not legalized (that is, the corresponding population is accounted for in the numerator but not in the denominator of the coverage ratio).

25. The concessionaire took over those utilities in 1998–99, but no data were available for the years before 2000.

the state utility Companhia de Saneamento do Parana (SANEPAR), whose coverage stood at 98 percent for water and went up from 31 percent to 61 percent for sewerage.

**Use of Hybrid PPP Schemes to Improve Access in Colombia**

More than 40 water PPPs have been awarded in Colombia since 1996, serving a combined population of 7.3 million people by 2007. Many contracts were awarded in poor municipalities with highly deteriorated infrastructure and relied on a mix of public and private funding. The first wave of contracts were awarded in 1996–98 and mostly followed the mixed-ownership company model, with the municipality holding a majority of the shares but with management fully delegated to the private operator. A second group of projects started in 2000 with the implementation of the Programa de Modernización de Empresas (PME) by the central government. The program focused on turning around public water utilities in very poor condition by transferring them to private operators, and featured concession schemes with significant public grants for investment.

Colombia's eight largest and/or oldest PPPs' performance in improving water coverage is compared with the national average improvement and with the achievements of the three largest public providers (see figure 3.2), with the period of comparison between the PPP's first year of operation and 2006. The eight PPPs reviewed represent a combined population of almost 4 million people and include all large contracts awarded before 2004.

The two mixed-ownership companies in Barranquilla (serving 1.3 million people) and Cartagena (serving 1 million) are the largest and oldest water PPPs in Colombia, and both have good records in expanding access. That in Barranquilla made notable progress in both water supply and sewerage services: coverage rose from 86 percent to 96 percent for water and from 70 percent to 93 percent for sewerage (1997–2006). Performance was even better in Cartagena, where water supply coverage jumped from 74 percent to almost universal coverage, while sewer coverage went up from 62 percent to 79 percent (1996–2006). Cartagena achieved full water supply coverage despite a 50 percent jump in the size of its population during the same period, largely due to the arrival of poor rural migrants. Half a million people gained access, and 60 percent of the new connections benefited families in the poorest income quintile. To achieve universal coverage, the operator in Cartagena made extensive use of community bulk-supply schemes that provided safe water to the many illegal settlements that were expanding on the city's periphery.

Mixed-ownership companies were also established in several mid-size Colombian cities. In **Santa Marta**, water coverage improved rapidly during the first three years, going from 74 percent to 87 percent, but it has stagnated since 2001 (access to sewerage followed the same pattern). In **Palmira** (220,000 people) and **Girardot** (100,000 people), full coverage was achieved for both water and sewerage.

The city of **Tunja** (120,000 people) was the first in Colombia to award a water concession. Both water and sewer coverage went up from 89 percent in 1996 at the start of the concession to 100 percent four years later. A series of smaller concession contracts were also awarded to national operators in 1997–98 for towns and small cities in the department of Antioquia. The largest of these operators is Conhydra, whose performance in expanding access appears to have been satisfactory. In the towns of **Marinilla, Santafe,** and **Puerto Berrio** (combined population 100,000), full coverage was achieved within a few years, starting from levels of 80–90 percent, and improvements in sewer access were even more

**Figure 3.2 Increases in Water Supply Coverage under Private Operators Compared with That of Public Utilities and with the National Urban Average in Colombia**

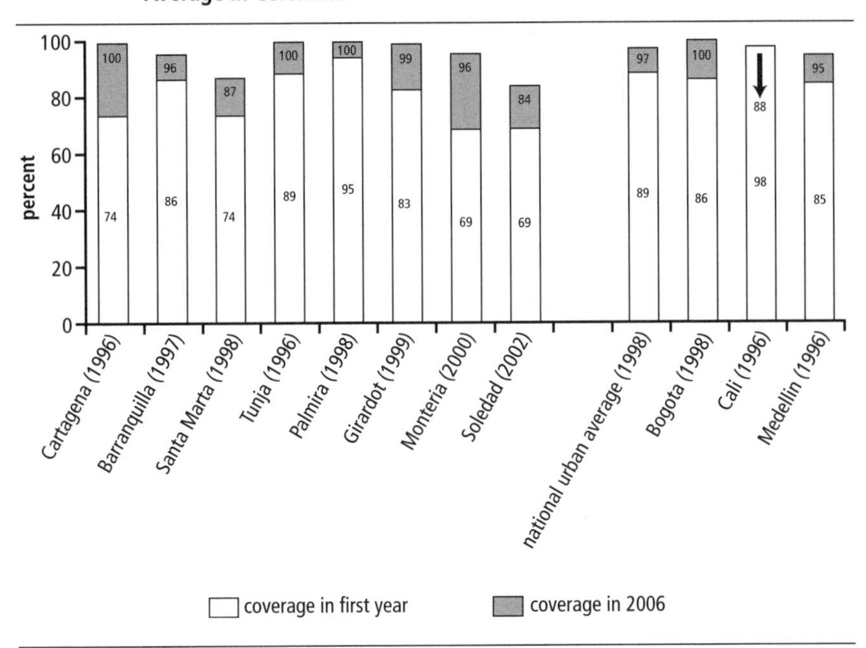


*Source:* National regulator (utilities data), with WHO/UNICEF and PAHO for national urban coverage.

*Note:* Coverage begins with the PPP's first year of operation and ends in 2006. The coverage figure in Cali is reported to have gone down from 98 percent to 88 percent between 1996 and 2006.

dramatic. All these concessions were designed around a mix of public and private financing, with municipalities contributing annual budget transfers from the central government.

The second phase of water PPPs in Colombia was shaped after 2000 by the implementation of the PME. Most of the cities concerned had high rates of poverty, and the schemes were based on concession contracts with public support for financing. The design was based on the central government providing grants in the start-up years to speed the rehabilitation of deteriorated systems and to expand access, while the contracting municipal government also made budgetary transfers on an annual basis to complement revenues.

The first contract under the PME was awarded in 2000 in **Monteria** (350,000 people), and its performance in improving water access has been very good. Starting from less than 70 percent, water coverage had jumped to 96 percent by 2007, catching up with the national urban average, and the population with access to piped water more than doubled. The improvement in sewer coverage was more modest, up from 26 percent to about 40 percent. In **Soledad** (400,000 people), coverage went up from 69 percent to 84 percent for water, and from 36 percent to 73 percent for sewerage, in just five years of operation.

Most of the PPP projects awarded under the PME were for small cities and towns, and limited data are available from Colombia's national database. Most contracts were signed between 2002 and 2004, and a comprehensive evaluation of the projects' performance has still to be carried out. Silva (2007) indicates, nonetheless, that coverage expansion under most PME projects appears to have been satisfactory.

Overall, water PPP projects in Colombia have performed well in expanding access. The largest and oldest PPP projects described earlier, which represent about half of the population served by private water operators, succeeded—in most cases—in improving access significantly, often in cities with high rates of poverty. The same holds true for those implemented under the PME since 2000. The financial approach adopted, which relied on a mix of private and public funding, no doubt contributed to their positive results.

However, it is not clear whether these partnerships performed significantly better than publicly managed utilities. The public utility in Bogotá, which serves one-third of Colombia's urban population, made strong progress in access over the past decade, reaching universal coverage. The performance of the public utility in Medellin was comparable, for expanding access, to the performance of the largest PPP in Barranquilla. This conclusion tends to be supported by two recent studies (Barrera and Olivera 2007; Gomez-

Lobo and Melendez 2007). Using household surveys instead of data from utilities, these studies did not find that private operators showed a clear advantage in improving water access, compared with public utilities. This can be attributed to several factors. The PME, which focused on introducing private operators into the worst-performing public utilities, must have helped to improve the average performance of the public sector.[26] A solid regulatory framework was put in place at the national level, and the recent development of local financial markets also allowed municipalities to access a viable source of investment funding for public water utilities, helping publicly managed utilities to compete on more equal terms with PPPs. Overall, the progress achieved in Colombia for access to water supply and sewerage services must be credited to several good performers in both the public and the private sectors and to a national policy that fosters accountability and efficiency of all service providers.

### *Large Concessions in East Asia*

The concessions in Manila (the Philippines) and Jakarta (Indonesia) have been in place now for almost a decade, serving a combined population of 18 million people in 2007. In both cities, the systems were divided into two separate concessions, a large portion of the population lacked access to piped water when the private concessionaires took over, and the first years of operation were severely affected by the 1997–98 Asian financial crisis.

#### The Two Concessions in Metropolitan Manila

Metropolitan Manila is the biggest metropolis in the developing world to be served by private water operators. Two concessions were awarded in 1997 to cover two contrasting areas. The Western concession (Maynilad, operated by Benpres-Suez) was the larger and covered the oldest and most developed portions of the city, with a total population of about 7 million people. The smaller Eastern concession (Manila Water, operated by Ayala-United Utilities) covered about 4 million residents and a large proportion of new neighborhoods in the process of development.

One-third of metropolitan Manila's population lacked access to piped water when the concessionaires took over. A key contractual target was the requirement to achieve universal access by 2006. The performance of the two concessionaires in expanding access is presented in figure 3.3, which

26. The impact of the PME was twofold. First, it removed the worst performers from the public utilities sample, since the corresponding utilities became private, thereby improving the average public performance. Second, the central government used the PME as a way to pressure nonperforming public utilities to reform.

shows the evolution of piped water coverage, number of connections, and number of households served for the concessions' first nine years.[27]

The implementation of the two concessions faced major difficulties from the beginning. The Asian financial crisis started just one month after the concessionaires took over. The Western zone operator, which had assumed most of the foreign currency–denominated debt of the previous public utility, found itself in virtual bankruptcy as the Filipino peso lost half of its value. Both concessionaires severely curtailed their investments during the first five years.

Though the contractual target of reaching full coverage by 2006 was not met, access to piped water in Manila expanded significantly during the decade. In the Western concession (Maynilad), coverage expanded from 67 percent to 86 percent. In the Eastern zone (Manila Water), coverage jumped from 49 percent to 94 percent. Meanwhile, the national urban average for water coverage grew moderately, from 46 percent in 1997 to 58 percent by 2004. An estimated 4 million people gained access to piped water in Manila during the period 1997–2006, about half of them as the result of low-cost community-based schemes, mostly in the Eastern zone. Access to sewerage services remained marginal, though, at about 10 percent by 2007.[28]

The evolution of water coverage has been markedly different between the two zones. In the Western zone, despite the financial difficulties, notable progress was made early on, but coverage stagnated after 2001 as Maynilad's financial and contractual situation deteriorated. Most of the expansion came through individual connections. In contrast, water coverage in the Eastern zone grew no faster than the national urban average until the 2003 rate rebasing, but subsequently, the concessionaire began to invest heavily in expanding its customer base. It installed as many as 160,000 new connections in just three years, between 2003 and 2006, and put in place a major program to expand access in poor areas through bulk-supply community schemes.

---

27. The coverage level that the concessionaires achieved remains the subject of much debate; the regulator has not yet ruled on the methodology to be used, quoting the concessionaires' estimates in its reports. These estimates are the figures quoted here.

28. The concessions were designed with a decision to expand access to sanitation not through a sewerage network but through individual septic tanks with the concessionaires being responsible for the periodic desludging of individual tanks. Between 1997 and 2006, sewer coverage went up marginally from 7 percent to 10 percent in the Eastern zone, and it went down from 14 percent to 10 percent in the Western zone.

**Figure 3.3 Evolution of Water Supply Coverage under PPPs in Manila, the Philippines, 1997–2006**

**(a) Private concessionaire coverage vs. national urban average**

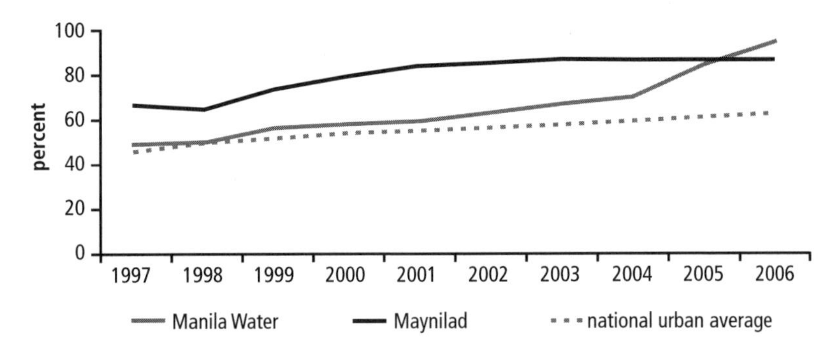


**(b) Connections and households served by concessionaires**

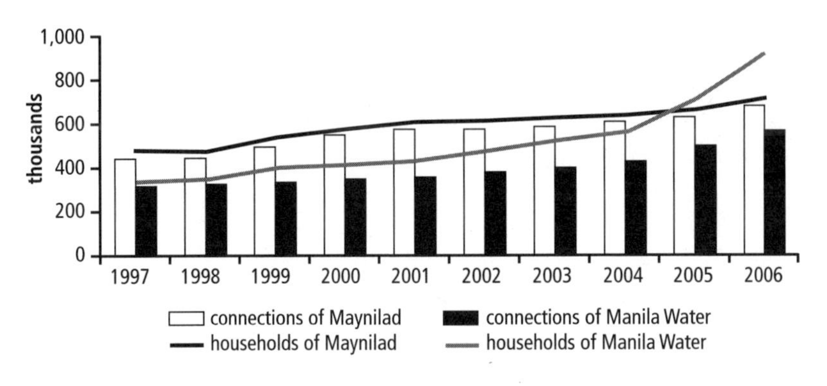


*Sources:* Regulator and companies' data.

## The Two Water Concessions in Jakarta

The two concessions in Jakarta were awarded in 1998, just one year after Manila's. Access was even lower, with water supply coverage at a mere 40 percent and a served population of about 4 million people. Many city dwellers obtained water from private wells.

Between 1998 and 2005, access to piped water in Jakarta's Western zone (operated by Suez) went up from 32 percent to 50 percent, while access went up from 57 percent to 67 percent in the Eastern zone (operated by Thames Water). Together, the two concessionaires added 210,000 water connections to the system in the first seven years, providing access for an additional

1.7 million people. More than 65 percent of the households that gained access through a household connection in the Western zone were registered as poor or very poor.

On balance, the concessionaires' performance in expanding access to water supply in Jakarta is rather mixed. It is positive when considering that significant progress was made in these two concessions, whereas nationwide urban water coverage stagnated at a low 30 percent in the same period. However, half of the population in the Western zone, and one-third in the Eastern zone, were still not connected to the water network after almost a decade. One reason is the widespread development of private wells in Jakarta and the absence of regulation and control over the use of the aquifer. Many households were unwilling to pay a connection fee plus periodic bills, because they had already invested in pumping equipment and were not paying any extraction charge for the water they were getting from their wells.

### *Experience in Africa*

Many water PPPs in Sub-Saharan Africa have followed the affermage model, often leveraging significant amounts of donor funding. Several affermages remained in place for more than five years, serving a combined urban population of more than 17 million (Côte d'Ivoire, Guinea, Niger, Senegal, and Maputo [Mozambique]) and providing a good sample to analyze performance. Another specifically African feature is the existence of mixed water and power utilities, with concessions put in place in Gabon and Mali for national utilities, and in Morocco for the cities of Casablanca, Rabat, Tangiers, and Tetouan (water distribution only).[29]

### Affermages and Combined Power and Water Concessions in Sub-Saharan Africa

Figure 3.4 shows the evolution of water access under private operators in Sub-Saharan Africa, measured in terms of household connections and improved access; the latter is an essential indicator, because a large portion of the urban population in the region relies on community standpipes.[30]

The private operator in **Côte d'Ivoire** (serving 7.5 million people) has been in place since 1960 and provides service to all urban centers and small towns in the country under an affermage arrangement. Between 1990 and 2006, improved access to piped water went up from 68 percent to

29. In Morocco, water, sewerage, and electricity services in large cities are provided by municipal utilities (either public or private) that focus on distribution of water and electricity, and that purchase water and electricity in bulk from national state providers (ONEP [Office National de l'Eau Potable] and ONE [Office National de l'Electricité], respectively).

30. The data and analysis presented here have been published in more detail in a companion study by Fall and others (2009).

**Figure 3.4 Evolution of Household Water Supply Coverage under PPPs in Sub-Saharan Africa**

**(a) Household water connection**

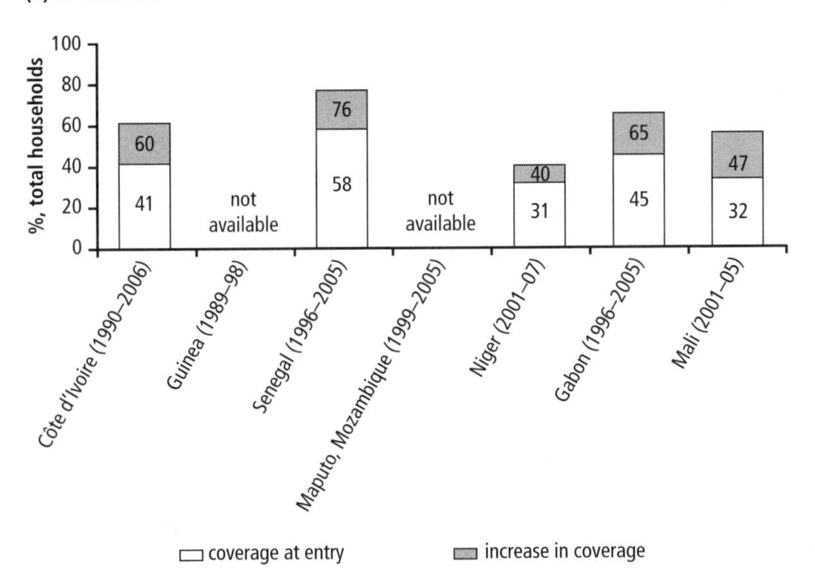


**(b) Improved access to water**

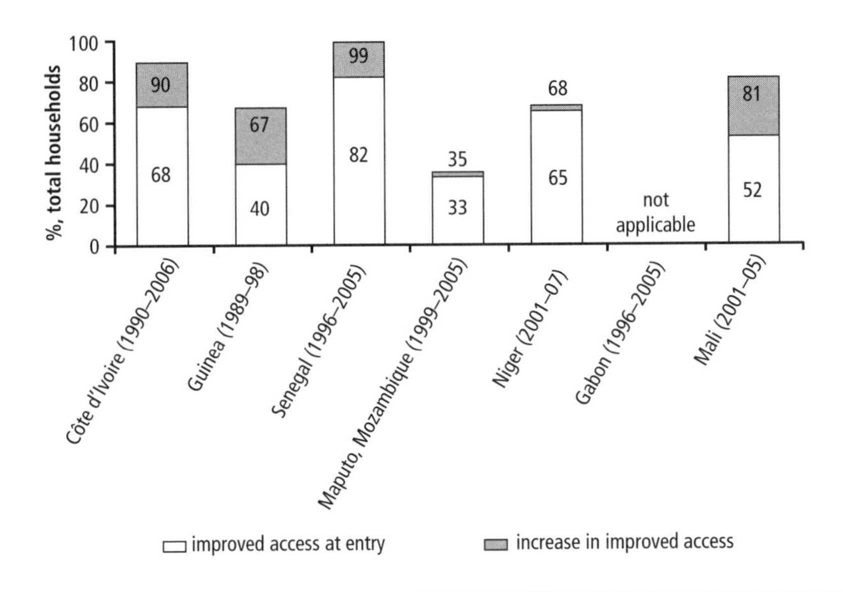


*Source:* Author; Fall and others 2009.

90 percent, and the number of people served more than doubled. Coverage through household connections went up from 41 percent to 60 percent, and the number of people served through residential connections also more than doubled, from 2 million to 4.9 million (see box 3.2). A total of 340,000 new water connections were installed.[31] This performance is even more notable when one considers that, unlike in a typical affermage (where most investment is financed by public money), almost no government or donor financing was provided during the past 15 years. A water development fund was established in 1988 to finance investment using a tariff surcharge, and the private operator is responsible for managing this fund and carrying out all civil works. Over the past 15 years, US$200 million was collected from customers through this fund and invested mostly in network expansion.

After Côte d'Ivoire, the next affermage implemented in Western Africa was signed in 1989 for the national water utility of **Guinea** (serving about 1 million people). During the first six years, the improved access ratio went up from 40 percent to 67 percent, with some 600,000 people (mostly in Conakry) gaining access to safe piped water. But after early progress, the PPP in Guinea encountered difficulties. Unlike the case of Côte d'Ivoire, in Guinea the civil works were not implemented by the private operator but by a newly established public asset-holding company responsible for investment. The coordination of civil works between the private operator and the public asset-holding company proved a continuous challenge, and major delays occurred with the investment program. Frustrated with the slow pace of the public asset-holding company's implementation of investments, the private operator directly arranged bilateral financing for civil works contracts, which were awarded to itself on a sole-source basis. Though this arrangement led to a timelier implementation of investments, it also distracted the operator from its main responsibility of running the utility and modified the incentives framework. The expected gains in operational efficiency failed to materialize, and the contract was not renewed after its expiration in 1998.

The **Senegal** affermage (serving 4.7 million people) started in 1996, and its design drew lessons from the Guinea experience by incorporating specific contractual targets and penalties to increase the operator's incentives to perform efficiently. The expansion of access in Senegal has been even more impressive than that in Côte d'Ivoire, thanks to an important injection of funds from donors and the cash-flow surplus generated by the operator and transferred to the asset-holding company. Unlike in Guinea, in Senegal the

31. To put this achievement in perspective for a poor country, one must consider that over a decade, private concessionaires in Buenos Aires and Manila installed 310,000 (Aguas Argentinas), 230,000 (Maynilad), and 250,000 (Manila Water) new connections.

Box 3.2

### Using Subsidized Water Connection Programs to Expand Access in Sub-Saharan Africa

A major element in the success of the PPPs in Côte d'Ivoire and Senegal for improving access through household connections was a social connection program to subsidize the cost of connection fees. In Senegal, connections were provided free of charge, and the beneficiaries had to pay only an advance deposit on consumption equivalent to 30 cubic meters.

In Côte d'Ivoire, the social connection program was funded through the Water Development Fund tariff surcharge. About 340,000 social connections were installed between 1990 and 2006, but many of the newly connected households proved unable to pay their quarterly water bills from their irregular incomes. The overall disconnection rate reached about 15 percent in 2002, and there were about 70,000 inactive connections in the country in 2006. Most of these disconnected customers are still accounted for as covered under the "improved access" criterion, because they obtain water from neighbors.

A social connection program was also implemented in Senegal, this time financed by loans from donors through the public asset-holding company. Some 129,000 connections (75 percent of all new connections installed) were installed under this program, benefiting poor households living in targeted neighborhoods. As in Côte d'Ivoire, a portion of the new connections were disconnected for nonpayment; most were families in the poorest income quintile. These experiences suggest that in the case of the poorest of the urban poor (who earn irregular wages in the informal economy) providing access to water through a household connection is not necessarily the most suitable solution, because many of these families find it difficult to save money for paying bills at the end of the month. In Niger, a more modest subsidized connection program (about 10,000 new connections) was also implemented during the first three years, funded through an International Development Association (IDA) credit and with the civil works directly implemented by the private operator.

public asset-holding company proved efficient in implementing the investment program to rehabilitate and expand the systems, and over a decade, improved access to piped water in urban areas jumped from 81 percent to almost full coverage. The number of residential customers went up from 217,000 to 375,000 people. The connection ratio went up from 58 percent to 76 percent and is now the highest in Western Africa.

In **Niger** (serving 1.8 million people), the affermage started in 2000, and its performance in improving access has been fair. The investment program during the first years of the PPP was more focused toward rehabilitation of existing assets than on expanding access. About 450,000 people gained access to piped water, but the coverage figures improved only marginally: the improved access and connection ratios went up from 65 percent to 68 percent and from 31 percent to 40 percent, respectively.[32]

The affermage in Mozambique's capital, **Maputo** (serving about 1 million people), also started in 2000. Its performance for expanding access so far has been disappointing, with coverage remaining broadly stable at a low 35 percent by 2005 (25 percent through residential connections and 10 percent through standpipes).[33] Difficulties have hampered the implementation of this PPP project since its inception. Catastrophic floods occurred in the first month after the takeover, making it necessary to divert some of the original donor funding to emergency repair works. The private operator left in 2001, and although it was replaced by another experienced foreign operator (its junior partner in the initial consortium), the renegotiation of the contract between the new parties was not finalized until 2004. Because the original design of the contract made the private operator directly responsible for implementing most of the civil works financed by donors, the result was serious delays in implementing the original investment program, which has been limited so far to rehabilitation and extension of production capacity instead of network expansion.

The performance of the two large concessions in Western and Central Africa for expanding access is satisfactory. In **Gabon** (serving 750,000 people), the concessionaire is responsible only for providing water services through household connections; standpipes remain the responsibility of a government agency. Since 1996, water supply coverage through connections went up from 45 percent to 65 percent nationwide; an estimated 300,000 people had gained access by 2007. In **Mali** (serving 1.6 million people), the private operator withdrew in 2005 but had significantly extended access to water supply during its four years of presence: improved access to piped

32. The fact that the increase in coverage figures is rather modest is partly due to fast urban population growth, because Niger has one of the highest fertility rates in the world.

33. Water access figures in Mozambique are not readily comparable to those from other countries of Sub-Saharan Africa. Reported government figures use a ratio of about five people per individual connection to report water coverage figures—a ratio that does not account for households purchasing piped water from neighbors. It is estimated that about a quarter of the residents in Maputo obtain piped water from their neighbors. In other Sub-Saharan African countries, a ratio of 8 to 10 people per individual connection is typically used when reporting access figures, to account for the widespread practice of purchasing piped water from neighbors.

water rose from 52 percent to 81 percent in the area covered by the contract, and about 600,000 people gained access (a 60 percent increase).

**Concessions for Water, Sewerage, and Power Distribution in Morocco**

The concession in **Casablanca** (serving 3.7 million people) has been in place for a decade. Its performance in expanding access has been satisfactory: between 1997 and 2005, more than 270,000 new water connections were installed, and about 1.3 million people gained access to piped water. Water supply coverage in the concession area went up from 71 percent to 93 percent in the first eight years. After taking over a utility in which coverage was 9 percentage points below the national urban average, the concessionaire was able to catch up to and even exceed this average by 2005.

The financial design of the Casablanca concession played an important role in supporting access expansion. A special work fund, financed by a 0.5 percent tariff surcharge, was set up; it financed US$140 million of the US$500 million of civil works that the concessionaire carried out over a decade. Another major funding source has been the special mechanism applicable to all of Morocco's water utilities (whether public or private), whereby new customers are required to pay a financial contribution well above the actual cost of connection.[34] Still, the operator did not meet targets for coverage expansion, mainly because of high connection fees and difficulties in dealing with illegal settlements.

No coverage data were obtained for the **Rabat** concession, but in **Tangiers** and **Tetouan** (together serving about 1.1 million people), water coverage rose from 67 percent to 76 percent and from 79 percent to 86 percent, respectively, between 2001 and 2005. These concessions had a financial design similar to that in Casablanca.

Data available for the 2001–05 period on the evolution of the number of customers suggest that the concessionaires in these four Moroccan cities performed well but not significantly better than public operators, using data from the four largest municipal utilities (Fez, Marrakech, Agadir, and Meknes). The number of customers grew at about 6 percent per year under the four concessions—a rate of growth similar to that achieved in Fez, Marrakech, and Meknes. The best performer during this period was the well-performing public utility in Agadir, where water coverage rose from

34. Under this scheme, a new customer must pay a financial contribution that is based on the property area and includes, in addition to the cost of connection, an amount equivalent to a portion of the value of the investment already made in existing systems. This financial contribution is called *participation aux frais de premier établissement* (participation in the costs of first establishment). For utilities located in fast-growing cities, it typically represents 10 percent to 15 percent of revenues.

66 percent to 81 percent, with an annual growth of 9 percent in the number of connections.

### *Conclusions on Access and Expansion of Coverage under PPPs*

Several conclusions can be drawn from this performance review of a large number of projects, regarding the overall contribution of public-private partnerships for expanding access in developing countries, the importance of designing sound financial arrangements, and paying sufficient consideration of the specific needs of the poor. Ultimately, many key factors for expanding access to basic water and sanitation services are beyond the scope of a utility, regardless of whether it is privately or publicly managed.

### More Than 24 Million People Gaining Access to Piped Water as a Result of Water PPPs

Many PPP projects failed to meet their contractual targets for coverage expansion, and often concessionaires failed to invest the amounts they had initially committed to expand the systems. Still, many PPP projects did achieve significant improvements in water and sewer coverage. This study found that since 1991, PPP projects in developing countries have provided access to piped water for more than 24 million people (see appendix B). Although this amount may be little in global terms, it is still significant, considering that private operators were serving less than 1 percent of the urban population in developing countries in 1997 and that their market share had risen to a mere 4 percent by 2002, reaching just about 7 percent by 2007.

The figure of 24 million people gaining access is quite conservative. It is based on 36 large long-term PPP projects (concessions, leases-affermages, and mixed-ownership companies) documented in this study. Together, these partnerships provided access to piped water to more than 24 million people, but there were more than 220 active water PPP projects in developing countries by the end of 2007. The estimate excludes the contributions of several large contracts that have been in place for more than five years but for which there were insufficient data (such as Aguascalientes, Campo Grande, Cancun, La Havana, Mendoza, Saltillo, and San Pedro Sula). Nor does it include the many PPP projects for towns and small cities in Argentina, Brazil, and Colombia. The figure also does not include the improvements achieved under management contracts, even though several of these achieved notable gains (as in Amman, where more than 400,000 people gained access).

### No Clear Advantage of PPPs for Expanding Access to Basic Services

Many water PPP projects have performed well in expanding access to piped water as well as sewerage services in some cases. Several projects stand out as very good performers, including those in Côte d'Ivoire and Senegal (national utilities), in the province of Corrientes (Argentina), in East Manila (the Philippines), and in the cities of Cartagena and Monteria (Colombia), Casablanca (Morocco), and Guayaquil (Ecuador).

Overall, though, there is no evidence that PPP projects are necessarily more efficient than publicly managed utilities for expanding access. In Argentina, Brazil, Colombia, and Morocco, for example, private concessionaires on average did not perform demonstrably better than public utilities, on the basis of data available. In Guayaquil (Ecuador), the concessionaire achieved remarkable progress in water coverage, but this can be largely attributed to a national public grant scheme that put the concessionaire at an advantage compared with other utilities in the country.[35] In Jakarta (Indonesia) and Manila (the Philippines), the concessionaires performed better than their counterparts in secondary cities, but they cannot be meaningfully compared with them because of the differences in size and conditions for access to financing. Finally, the performance of PPP projects in expanding access to sewerage has been uneven.

In Sub-Saharan Africa, private operators have clearly performed better than public utilities in expanding access through household connections. PPP projects accounted for almost 20 percent of the increase in household connections in the region, more than twice the amount that would have been predicted on the basis of their 9 percent market share.[36] But half of these gains were made in just one country (Côte d'Ivoire), and when using the improved access criterion—which also considers more basic forms of access (such as community fountains) and one used for tracking whether or not countries meet the Millennium Development Goals (MDGs)—the difference between public and private national utilities is not obvious (Fall and others 2009).

### The Close Link between PPP Coverage Expansion and Financing

The performance of PPPs in coverage expansion is closely linked to financing. Expanding access to piped water and sewerage services usually entails major investments. Not surprisingly, much of the diversity in performance

35. The proceeds of the telephone tax are allocated among utilities, on the basis of where the call originated. Because Guayaquil is the largest city and the economic capital, it gets the lion's share of the proceeds. Under the previous public utility in Guayaquil, the telephone tax was already in place but raised much less money, because the telecom boom took place in the early 2000s, just when the concession started. Still, the private operator proved efficient in making use of the opportunity given by this mechanism to rapidly expand its customer base, benefiting at the same time the many poor who thereby gained access to safe piped water.

36. Throughout Sub-Saharan Africa, 27 million people have gained access to household connections since 1990, and although water PPPs served only 9 percent of the urban population, they provided more than 5 million people with household water connections (3 million in Côte d'Ivoire, 1.5 million in Senegal, and 0.3 million in Gabon).

among PPP projects can be traced to differences in financial design and availability of funds for investment.

In concessions, the performance in expanding access was greatly influenced by the financial conditions of the contract, which affected how much investment could be financed through cash-flow generation and whether and how much the concessionaire could borrow. The assumption that private lenders would be ready to provide large amounts of nonrecourse project financing to private operators proved unrealistic. Several concessions proved vulnerable to economic crises, especially when concessionaires had borrowed heavily in foreign currencies. In Manila, the performance in early years was seriously affected by the Asian financial crisis, and the Eastern zone concession started to make progress in extending access only after the regulator had granted a tariff adjustment—seven years after the start of the project.

In affermages, the government has the largest role for financing investments, and how the investment program is executed has played a major role in the overall outcome of the PPP projects. Various situations are encountered in practice regarding who actually implements the civil works. In many affermages in Sub-Saharan Africa, investment has been carried out by a public asset-holding company, and the outcome for access expansion has depended on government priorities and on how fast civil works were implemented. In Senegal, the government adopted access expansion as a priority and, thanks to the good performance of both the asset-holding company and the private operator, has achieved the best access figures in the region. In Maputo (Mozambique), delays in completing the investment program, plus the priority given so far to rehabilitation and expansion of production capacity, have hampered progress in expanding coverage. Côte d'Ivoire is a special case because it is more of a hybrid between affermage and concession: investment has been financed entirely through cash-flow generation, and the private operator implemented the civil works directly. In mixed-ownership companies under a lease arrangement such as in Cartagena (Colombia), the utility under private management carried out all civil works, but investment decisions were made jointly by the private partner and the city government.

Notably, the design of many of the successful PPP cases included the provision of public financing (through grants or concessional loans) to accelerate access expansion. The advantages of such an approach are apparent from the good performance of the concessions in Cordoba (Argentina) and Guayaquil (Ecuador) and those awarded in Colombia under the PME (where a significant amount of public funding was provided for investment). By contrast, the difficulties in La Paz–El Alto (Bolivia) and the Buenos Aires province (Aguas de Bilbao [Argentina]) concessions show the limitations of relying solely on the sector's own cash flow to finance expansion when a large portion of the population is poor. As for affermages, the fact that the water coverage achieved in Senegal (where a major injection of donor funding took place) is much greater than that in Côte d'Ivoire (where all expansion was financed through customer revenues) shows that the level of achievement depends heavily on how much a government is willing to contribute for meeting social goals.

**The Importance of Providing Low-Cost Service Alternatives**

Providing low-cost alternatives adapted to the needs of the poor is important. In many developing countries, affordability is a significant issue for poor families, for whom the cost of connecting to the network can be prohibitive. Several private operators have offered financing schemes to help poor households pay connection fees (as in Argentina, La Paz–El Alto in Bolivia, Colombia, and Manila in the Philippines), but this is not always enough. Subsidized connection programs in Côte d'Ivoire and Senegal have played a major role in the good performance of these two PPP projects, though the significant rates of disconnection suggest that household connections are not necessarily the most suitable service option for the poorest urban families.

Several of the most successful PPP projects stand out for having developed alternative, lower-cost service options through community-based schemes. Examples are Cartagena and Eastern Manila, both of which had achieved almost universal access to piped water by 2007 despite high poverty rates. In Cartagena, the expansion in access was part of a pragmatic, phased approach whereby piped water was provided in bulk at the entrance to each settlement in a first phase, while tertiary network and standard connections were gradually installed later as each settlement became legalized by the municipality. Though the service initially provided was less convenient than an individual connection, it still represented a major improvement over the previous situation and allowed a larger number of poor people to rapidly gain access to safe piped water.

It is also increasingly recognized that the small local private operators, who in many large cities of the developing world provide service to poor periurban neighborhoods, can play an important role in spearheading progress in expanding access for the urban poor. These largely informal players have usually surged in response to the deficiencies of the main water utility and, if properly regulated, could become valuable partners for improving access for the poor. A noteworthy feature of the large investment program financed by donors in Maputo (Mozambique) is that it is not limited to the lease contract with an international operator, but also includes financial support for small local operators to help them improve service quality and access.

**Factors That Are Beyond Utilities' Scope for Expanding Access**

Several key factors for expanding access go beyond the scope of a water utility—whether it is publicly or privately managed. Illegal settlements and slums house a high portion of the population in large cities of the developing world, but water utilities are typically ill equipped to serve these neighborhoods. Expanding networks in illegal settlements is often prohibited by law and raises land property issues that only active collaboration with local authorities can solve. Relying on community schemes for rapidly expanding access, as was done in Cartagena and East Manila, is not always legally possible because of prevailing technical standards, or accept as a viable solution by the population.

The fact that households are not always willing to connect to the water supply or sewerage networks can also be a challenge. In Jakarta, part of the reason coverage increased only modestly was that many residents already obtained water from wells and were often not interested in connecting to the water network. Even though this situation is creating problems for the environment (the current overexploitation of the aquifer) and public health (most of the water from wells is not safe), achieving better coverage figures would have required the government to enforce more control over the use of the aquifer. The same difficulty occurs when municipalities or utilities try to get households to connect to a newly installed sewerage network when they have already invested in individual sanitation. These are situations that a utility cannot solve unless the government is willing to enforce strict rules for making connection compulsory and/or to subsidize connection costs for new customers.

## Quality of Service

Quality of service can take various forms. This study focused on the performance of water PPP projects for improving the continuity of water distribution (when a private operator took over a utility with intermittent service and water rationing) and for complying with drinking water standards.

### *Reductions in Water Rationing*

Many water utilities in the developing world struggle with water rationing and low water pressure, usually because of massive water leaks in highly deteriorated networks. The poor are disproportionately affected. They tend to live in periurban areas on the edge of the distribution networks, where service pressure is lower and water often arrives only for a few hours during the night. And they are less able to afford equipment, such as rooftop water tanks and filters, to mitigate the negative impact of intermittent supply.

The ability of a water utility to provide an uninterrupted supply of piped water is probably the most important determinant of the quality of service. Without service continuity, potability cannot be guaranteed because of the risk of external infiltration and contamination in pipes. Water rationing lies at the root of the vicious circle of poor maintenance and service. Once rationing has become an established operating pattern, a network deteriorates faster because of repeated pressure surges. Attempts to restore continuous service often fail because any temporary gain in average pressure causes more pipe ruptures and seal failures, and hence, more water is lost to leaks. Because of this, utilities with deteriorated networks often adopt the short-term solution of reducing the number of hours of service to limit leaks, even though this practice produces poorer service to customers.

Reestablishing continuous supply once water rationing has been in place for many years is very difficult. Some public utilities in the developing world have succeeded in doing it, as in the cases of Phnom Penh (Cambodia) and Ouagadougou (Burkina Faso), both of which received support from international financial institutions. The review focuses on those PPP projects that started under conditions of water rationing and for which reliable data were available on the evolution of the average number of hours of service.

**Reductions in Water Rationing through Concessions and Leases-Affermages**

The widest body of evidence on the performance of long-term PPPs to reduce water rationing comes from Colombia, a country that offers a unique set of conditions for evaluating whether PPP projects can be efficient in reducing water rationing: first, rationing there is a widespread problem; second, private operators were introduced in the worst-performing utilities; and third, data on service continuity are available from the regulator. Figure 3.5 shows the evolution of the average number of hours of service in 10 PPP projects in Colombia that started under conditions of water rationing. In all cases, strong progress was made, and service continuity was often reestablished after five or six years. In the case of the PPP projects implemented under the Programa de Modernización de Empresas (PME), the concessionaires started with more severe rationing but benefited from public grants to spearhead rehabilitation. The good performance of private operators for reducing water rationing in Colombia was confirmed by findings from national household surveys (Barrera and Olivera 2007; Gomez-Lobo and Melendez 2007).

**Figure 3.5 Evolution of Service Continuity in Selected PPPs in Colombia**

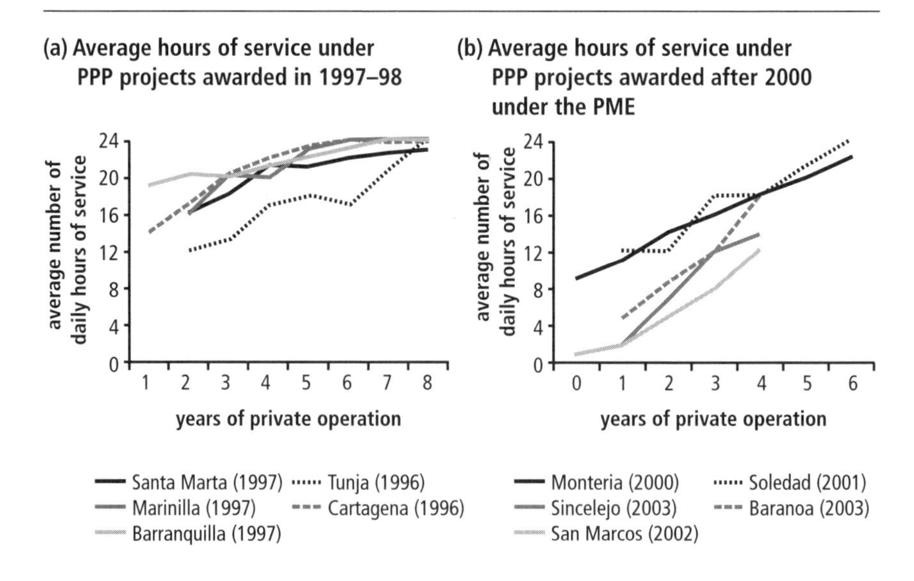


*Source:* National regulator.

*Note:* The year following the city or town indicates the year the PPP project began operation. For the reader's ease, the cases of Palmira (1998) and Girardot (1999) are not shown. In both these cases, water rationing was not initially as severe as in other cities and continuous service was reestablished in less than three years.

Apart from Colombia, few data are available from other Latin American countries on service continuity for long-term PPP projects. Many public utilities were providing continuous service before being transferred to private operators (as in Chile, but also in La Paz, Bolivia's capital). In Buenos Aires (Argentina), water shortages that had plagued the population during summer months were ended in the first year the concessionaire took over (1993), and the overall improvement in service continuity was sustained over several years (Delfino, Casarín, and Delfino 2007).[37] In Salta (Argentina), the proportion of the population suffering intermittent service fell from 43 percent in 1998 to less than 10 percent by 2006. In Guayaquil (Ecuador), half of the population suffered from rationing when the concessionaire took over in 2000, and little improvement had been made by 2005 (Yepes 2007).

37. Delfino, Casarín, and Delfino (2007) mention that the proportion of customers with appropriate water pressure jumped from 17 percent in 1993 to 60 percent in 1998, and to 74 percent in 2003.

In Western Africa, the overall experience with continuity of service has been positive (Fall and others 2009). In Dakar (Senegal), the private operator started with an average of 16 hours per day, and continuous service was reestablished by 2006. The fact that it took a decade to achieve this result—after combining investments in production capacity by the public asset-holding company with leak reduction by the private operator—underlines the difficulty of ending water rationing in a large and complex system while simultaneously expanding the population served. In Conakry (Guinea), continuous supply was reestablished within a few years of private operation but deteriorated again after the contract was concluded in 1998. In Niger, gradual progress was achieved in the capital, Niamey, from 2000 to 2006, improving from an average of 18 hours per day to 21 hours.

Few data were available from other regions. In Asia, the two concessions in Manila (the Philippines) present a contrasting picture. The Eastern zone concessionaire started with a very poor situation in 1996, with about 75 percent of its customers affected by water rationing. After a decade, it had reestablished continuous service for the whole concession area by 2006. In the Western zone concession, about 80 percent of the population received continuous service by 2001, but the situation rapidly deteriorated afterward as the concessionaire slid into bankruptcy; half of the customers suffered from intermittent service by 2005 when the contract was terminated. In Turkey, the lease contract in Antalya (serving 0.6 million) achieved a sizable reduction in rationing during its five years of operation (from 16 to 21 hours of service per day on average), but the contract was terminated in 2002 following conflict between the parties.

These achievements should be highlighted in the cases of Senegal; East Manila (the Philippines); and the Colombian cities of Cartagena, Barranquilla, Monteria, and Soledad. Not only was full continuous service reestablished after starting from significant rationing, but as was shown previously, these PPP projects simultaneously achieved very significant increases in water coverage. If the Millennium Development Goal criteria for access to water were taken literally—by measuring coverage as access to safe water (that is, with continuous service) instead of just possession of a connection—the net improvements in access would be even greater.

**Effectiveness of Management Contracts in Reducing Water Rationing**

Another sample set is provided by management contracts, a significant number of which were implemented in situations of water rationing. In these contracts, the evolution of the number of hours of service was often tracked closely, being one of the contractual targets used to determine the private operator's remuneration. The performance of the 12 management contracts for which data were available is shown in figure 3.6. The figure compares

**Figure 3.6 Gains in Service Continuity under 12 Management Contracts**

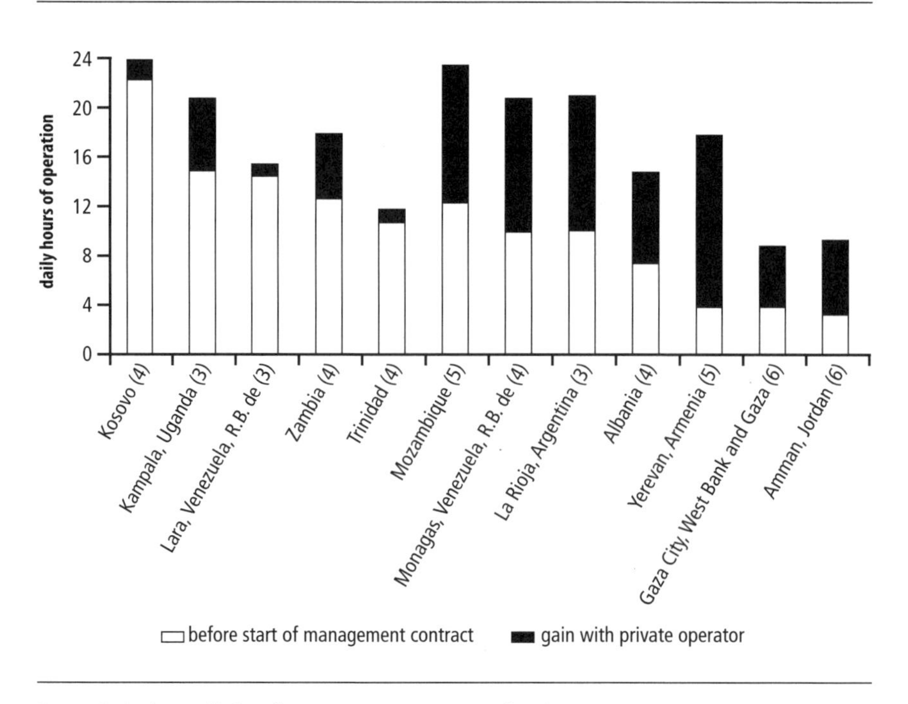


*Source:* Author's compilation of company or government project data.
*Note:* Years of private operation are indicated in parenthesis.

the average number of hours of service before the entry of the private operator with the level achieved by the contract end.

Although the methodology used for measuring the number of hours of service varies widely across contracts, a rather consistent pattern appears. Out of the 12 documented PPP projects that started with intermittent service and for which reliable before-and-after data were available, in 10 cases water rationing was significantly reduced by the end of the contract. Progress was particularly significant in Mozambique,[38] Monagas (República

38. The management contract in Mozambique covered four cities serving a combined population of about half a million people (the same operator also runs the capital, Maputo, under a lease contract). Progress was particularly notable in the cities of Beira and Quelimane, where full, uninterrupted service had been reestablished by the end of the contract in early 2008, starting from an average of less than 10 hours per day before the PPP.

Bolivariana de Venezuela), La Rioja (Argentina),[39] and Yerevan (Armenia). Only in a few cases was no significant improvement achieved: Trinidad and the Venezuelan state of Lara, to which should be added the cases of Chad and Guyana, which are not shown in figure 3.6 for lack of reliable data.[40]

Two cases deserve special notice. First, **Yerevan** (Armenia) typifies the situation where water rationing results from a combination of high leaks and customers' waste because of the absence of metering. When Yerevan's management contract started in 2000, water was supplied for about six hours per day on average. The contract was designed so that the private operator would have access to a rehabilitation fund, with flexibility to select and directly carry out priority civil works. Once the government had passed a law to aid in the installation of water meters for residential customers, a comprehensive installation campaign was carried out, together with network rehabilitation and repairs of buildings' plumbing systems to reduce water losses. The continuity of service increased to 18 hours per day, exceeding the contract target by more than 25 percent, and by 2005, 70 percent of the population enjoyed continuous service.

Second, **Uganda's** experience emphasizes that differences in system size prevent meaningful comparisons of performance between systems. The private operator improved the continuity of service in the capital, Kampala, but faster progress was made in smaller towns, which had remained under public management (Mugisha and others 2007). This can largely be explained by the fact that smaller towns have distribution systems that work on simple hydraulics and are easier to fix, something also attested to by the experiences of management contracts in Mozambique and Albania. In each of those countries, one private operator is operating in several cities of various sizes, and a much greater reduction in rationing could be achieved in small cities than in the two larger ones (Durres [Albania] and Maputo [Mozambique]).

### *Improved Compliance with Drinking Water Standards*

The impact of PPP projects on the quality of water delivered is harder to assess than the impact on water rationing. The concept of potability involves the compliance of water samples with multiple chemical parameters. The methods used for water sampling (frequency, number, and relevance of sampling points) can vary widely and have a major impact on the results. Reliable baseline data are typically not available, because many of the utilities

39. The PPP in La Rioja started as a management contract and evolved into a concession after year three.

40. For the two PPPs for the national utilities in Chad and Guyana, no reliable before-and-after data were available, but available information suggests that little progress, if any, was achieved in reducing water rationing as a result of the management contracts.

transferred to private operators had not been previously conducting proper sampling and analysis. Meaningful project-specific data on the evolution of water potability in PPP projects are, therefore, hard to come by.

The few econometric studies that have looked at the impact of PPP projects on the quality of water delivered to customers all point to a clear positive impact. Andrés, Guasch, and others (2008) found that water potability in Latin America improved significantly with the introduction of a private operator, in both the transition and the post-transition periods. In Colombia, using household and public health surveys, both Barrera and Olivera (2007) and Gomez-Lobo and Melendez (2007) found that PPP projects tended to achieve better potability figures than did public water utilities. In Argentina, Galiani, Gertler, and Schargrodsky (2005) found that the child mortality rate fell in areas served by private operators.

The few data available in Latin America come from Argentina. **Buenos Aires** is one of the rare PPP projects for which reliable data are available on the yearly evolution of water potability for three key potability parameters (turbidity, chlorine, and bacteriology). As for access and service continuity, the concessionaire performed well in early years. Before the concession began operating, only half of the water samples in Buenos Aires complied with turbidity standards, one-third had insufficient chlorine, and almost 10 percent tested positive for fecal contamination. An overall compliance rate of more than 99 percent was achieved by the fourth year of private operation. However, Ducci (2007) reports that problems with water quality compliance started to occur after 2002 in Buenos Aires and also in the Santa Fe province concession. In Salta, the concessionaire has achieved gradual and consistent improvement in overall water potability in the various systems it has been operating in the province since 1998 (Yepes 2007).

In **Manila** (the Philippines), significant improvements were achieved by both concessionaires after they took over from public management. In four years (1996–2000), the level of potability compliance went up from about 96 percent to almost 100 percent, after a more stringent system of water quality monitoring was put in place under the control of the regulator. These improvements were sustained in later years, even in the Western zone, despite the concession going gradually into bankruptcy. Potability compliance in both concessions has averaged about 99 percent for the past seven years.

Positive performance was also registered for several PPPs in **Western Africa.** In Senegal and Niger, the improvement in water potability broadly followed the progress with service continuity. Potability compliance in Dakar went up from 95 percent in 1997 to 98 percent by 2001, and in Niamey it went from 96 percent to 98 percent in the first four years. In Gabon, potability also improved after the start of the concession, with the average turbidity index in Libreville falling from 2.5 to below 1.0. Finally, the population in Abidjan (Côte d'Ivoire) has enjoyed safe tap water for decades, a notable exception in the subregion.

## Operational Efficiency

The analysis required to fully assess the operational efficiency of a water utility—whether publicly or privately managed—is very complex, and undertaking individual efficiency analyses for a comprehensive set of PPP projects would have obviously exceeded the scope of this study. The cost structure of a water utility is made up of many factors, and efficiency gains can be achieved through different dimensions (such as changes in labor, leak reduction, or better use of chemicals or electricity) that involve multiple parameters. In most cases, information constraints limit the scope of analysis; for instance, analyzing the impact of energy and chemical use on a utility's efficiency cannot be done without disaggregated cost data, which are seldom available.

In practice, though, a large portion of a water supply company's operational efficiency can be captured by three key indicators—water losses, bill collection, and labor productivity—as follows:

- Water losses are a key cost element in most water utilities in developing countries. The non-revenue water (NRW) ratio is measured as the difference between the volume of water produced and of water billed to customers, divided by the volume of water produced. This ratio captures the efficiency both of the distribution network (physical losses) and of commercial management (commercial losses due to metering and billing problems). The evolution of NRW is usually a good proxy for variable costs.
- The bill collection ratio directly affects the cash flow of the utility and captures a large portion of the efficiency of commercial management.
- Labor productivity is a major input into an analysis of efficiency, labor being usually the largest fixed cost for a water utility.

These three indicators are discussed here in turn. The analysis of operational efficiency is able to draw on a larger sample of PPPs than the analyses of access and quality could. Among long-term PPPs (divestitures, concessions, leases-affermages, and mixed-ownership companies), 49 projects were reviewed, representing a combined population of more than 82 million people. A sample of 17 management contracts was reviewed, representing a combined population of more than 15 million people. These two samples combined represent close to 80 percent of the population served by long-term PPP projects signed before 2003 (excluding Eastern Europe) and management contracts signed before 2005.

### *Reductions in Water Losses*

High water losses are a widespread problem in developing countries. They are made up of two elements—physical losses caused by leaks in the distribution network, and commercial losses corresponding to water effectively delivered but not billed—both of which add to the operational costs of a water utility but are of a very different nature. A recent World Bank study estimated the full cost of water losses from urban water utilities in developing countries to be as much as US$5 billion per year (Kingdom, Liemberger, and Marin 2006).

Among published econometric studies on developing countries, only Andrés, Guasch, and others (2008) and Gassner, Popov, and Pushak (2008a) covered the impact of PPPs on water losses. Both studies found that the entry of a private operator significantly reduced water losses.

Project-specific data on water losses can be difficult to analyze. Though physical and commercial losses differ and have different remedies, reliable segregated data are usually not available. The indicator most widely used to discuss the performance of a water utility with regard to water losses—the NRW percentage—has limitations in practice and is not always the most appropriate indicator to assess the efficiency of a distribution network (Kingdom, Liemberger, and Marin 2006).[41] In general, data on water losses, and especially the baseline level when the private operator takes over, tend to be unreliable.[42] Furthermore, in many countries, households are billed on the basis of consumption estimates (because of either low metering coverage or malfunctioning meters), making it very difficult to assess the actual level of water losses. Nonetheless, a clear picture can be drawn by looking at a large number of projects in various countries and regions. This study first considers long-term PPPs in various countries and regions, and then analyzes the record of management contracts separately.

41. Physical losses in a network are largely driven by the number of connections (a major source of leaks at pipe junctions), the overall pipe length, and the service pressure. Because both the number of connections and the network length are key structural factors, the International Water Association (IWA) has been recommending complementing the widely used NRW percentage indicator with the average daily volume of water losses per connection or per kilometer of network.

42. The public utilities that were transferred to private operators often lacked the proper framework to calculate the level of water losses (owing to the absence of reliable macro-metering at production facilities, poor operational monitoring, and problems with the customer database, among other causes). In many cases, the baseline NRW level that had been estimated during the tender process was revealed to be grossly underestimated after the private operator had taken over and started to implement a proper measurement system.

**Reductions in Water Losses: The Case of Colombia**

Colombia is one of the few countries in Latin America where billing of residential customers is largely based on actual metering instead of consumption estimates, and the national regulator's database contains a large sample of PPP projects. The evolution of the NRW indicator for the eight largest and/or oldest PPP projects, which represent more than half the Colombian population served by private operators, is shown in figure 3.7. The NRW level in the year when the private operator took over is compared with the level achieved in the last year of available data.

The assessment based on the NRW indicator is quite mixed. Strong gains were made in Monteria, Tunja, and Palmira, but the reduction seems to have been more modest in Cartagena, Barranquilla, and Santa Marta, and no progress was achieved in Girardot or Soledad. However, using only the NRW percentage to gauge the evolution of water losses can be misleading in situations where major changes are made in the distribution network. This is the case especially when moving from intermittent to continuous service and expanding coverage—which is, indeed, what happened in the largest PPP projects in Colombia, as was discussed. For the three largest and oldest PPP projects, using the alternative indicator of water lost per connection provides a clearer picture of the performance for reducing water losses because it takes into account the major structural changes that took place due to system expansion. It shows that losses were more than halved in Cartagena and Barranquilla, and were reduced by 40 percent in Santa Marta. This was achieved while the average network pressure went up significantly as service continuity was reestablished—emphasizing that major improvements were made in the hydraulics of the distribution networks.

**Good Performance of PPPs in Reducing Water Losses in Morocco**

Large municipal water utilities in Morocco concentrate on the distribution and commercial functions, purchasing water in bulk from the national utility ONEP. Reliable data are available from the central government on the performance of both public and private providers.

The analysis compares the performance of four private concessions (Casablanca, Rabat, Tangiers, and Tetouan) with that of the six larger municipal utilities (Marrakech, Fez, Agadir, Meknes, Kenitra, and Oujda) in reducing non-revenue water. Figure 3.8 shows that the four city water systems operated by private concessionaires all achieved a marked reduction in NRW.[43] In the public utilities sample, Agadir stands out as an excellent

43. In Rabat, the level of NRW had initially deteriorated under the first private operator, which had taken over in 1999. The data in the figure represent the evolution under the current operator, which replaced it in 2002.

**Figure 3.7 Evolution of Water Losses under Eight PPPs, by NRW Level and Losses per Connections in Colombia**

**(a) Changes in NRW percentage indicator**

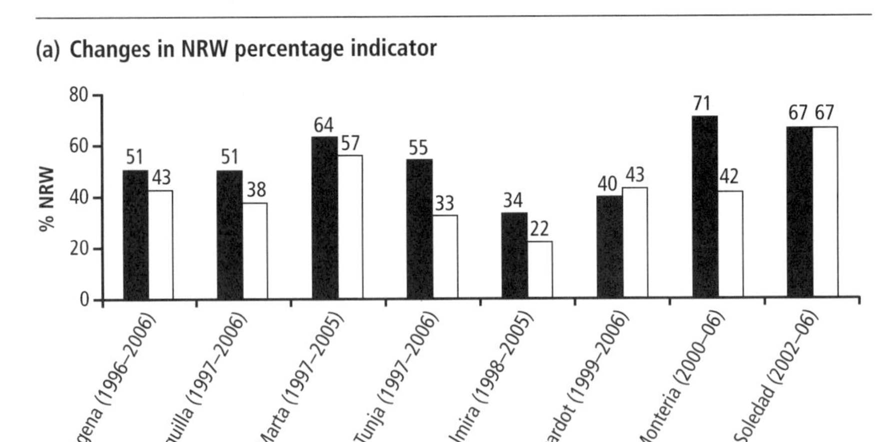


**(b) Daily losses per connection**

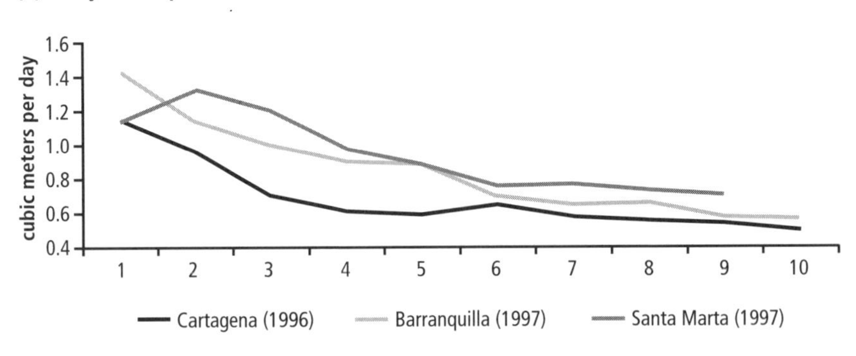


*Source:* National regulator database.

performer. Among the other five public utilities, Fez's was the only one to show an improvement over the past four years, but this improvement was much smaller than in the Tetouan concession, which started at a similarly high level of loss.

The reduction in water losses achieved by concessionaires in Morocco is even more evident from the evolution of the losses-per-connection

**Figure 3.8 Water Losses under Private Operators and Public Utilities, by NRW Level in Morocco**

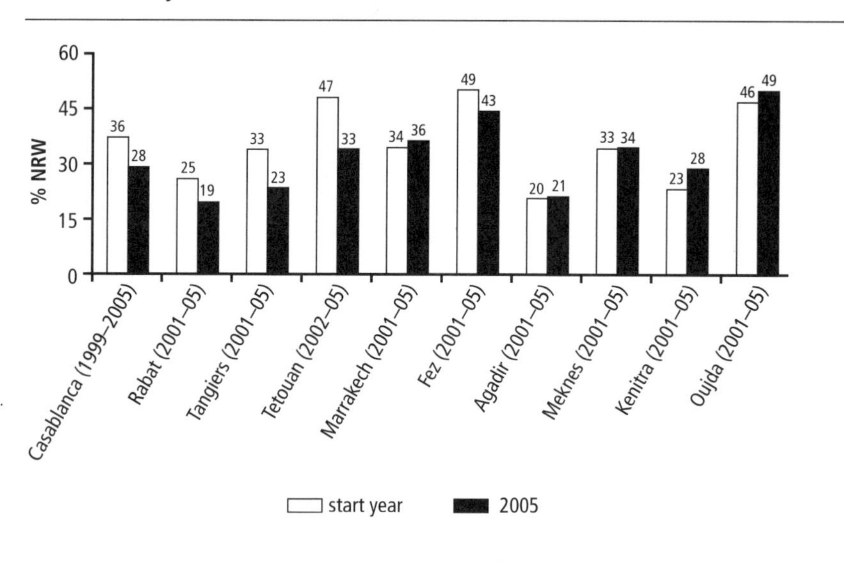


*Source:* Morocco Ministry of Interior, Direction des Régies et Services Concédés (DRSC).

indicator. The evolution of water losses in cubic meters per connection per day is shown in figure 3.9 for the same sample, putting public and private utilities side by side using the same vertical scale. The concessions in Casablanca and Rabat were able to catch up with Agadir's, the most efficient public utility, and the concessions in Tangiers and Tetouan were performing better than any of the other public utilities except Agadir's by the end of 2005.

### Reductions in Water Losses in Sub-Saharan Africa

The performance of most of the Sub-Saharan African concessions and leases-affermages that have at least two years of operation is shown in figure 3.10. Together, these PPP projects serve about 18 million people.

Most of these PPP projects significantly reduced water losses. PPP projects in Gabon, Niger, and Senegal have achieved NRW levels below 20 percent, comparable with those in well-managed utilities in Western Europe and North America. In Côte d'Ivoire, where the private operator has been in place for more than four decades, NRW went up from 15 percent to 23 percent between 1989 and 2006, but water losses per connection

**Figure 3.9 Evolution of Water Losses under Private Operators and Public Utilities, by Connections in Morocco**

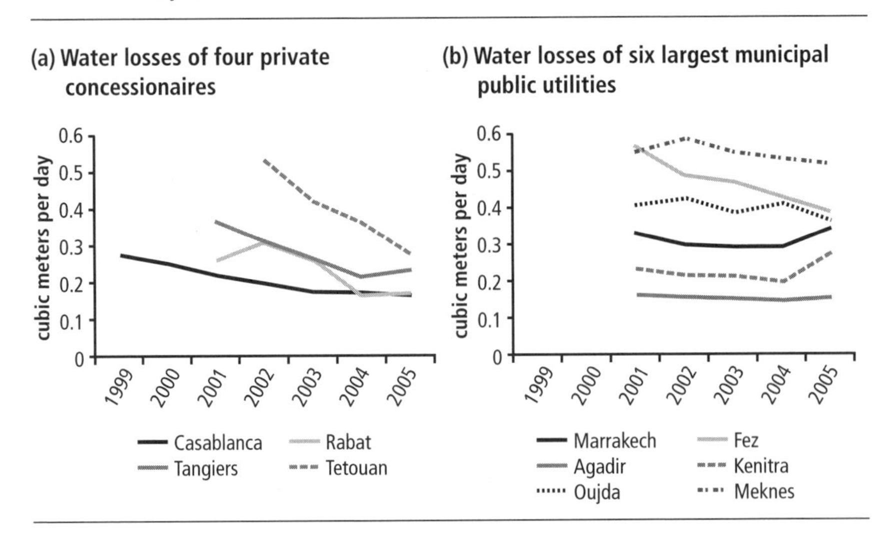


*Source:* Morocco Ministry of Interior, Direction des Régies et Services Concédés (DRSC).

**Figure 3.10 Water Losses under Eight Long-Term PPPs, by NRW Level in Sub-Saharan Africa**

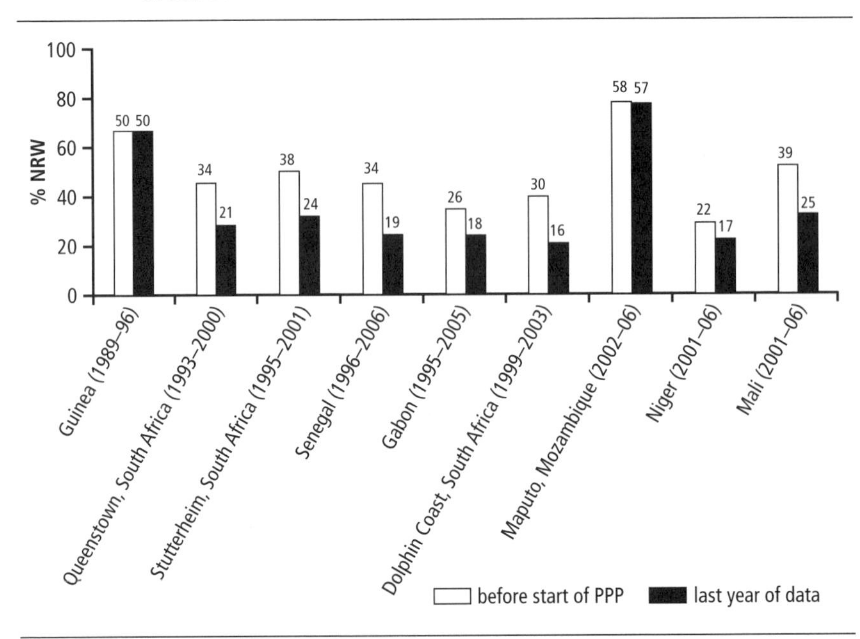


*Source:* Author's calculations based on various sources (see appendix A).

*Note:* Projects are listed in order of the year they began; data years are in parenthesis.

remained stable, at 0.18 cubic meters per day. In South Africa, PPP projects achieved notable NRW reductions in the three cases for which data were available (Palmer Development Group 2003).[44]

Guinea and Maputo (Mozambique) stand out as two cases where a lease-affermage contract failed to reduce NRW levels after several years of private operation. In Guinea, this can be linked to difficulties in the implementation of the investment program by the public asset-holding company, as well as a lack of contractual incentives for the private operator—a feature that was corrected in the design of the subsequent affermage in Senegal (box 3.3). In Maputo, the PPP project had experienced difficulties since its inception, and the implementation of the rehabilitation program was seriously delayed. Dealing with a high level of commercial losses has been a special challenge: less than half of the residential customers are metered, and many sell large volumes of water to their neighbors but are billed for only 10 cubic meters per month.

**Reductions in Water Losses in Latin America**

The performance of private operators in reducing water losses has been quite diverse across Latin American countries. The case of Colombia was already presented. Elsewhere in the region it is not uncommon for utilities to rely—at least partly—on estimates to bill households, making it difficult to track the actual evolution of water losses. This is especially the case in Argentina, where households often can choose whether to have their consumption metered, with unmetered customers billed using formulas based on housing characteristics that have no link to actual consumption. In Buenos Aires (Argentina) for instance, only 12 percent of residential customers had meters in 1998, so the NRW figure is not representative—although in this case, some evidence suggests that the concessionaire did reduce leaks.[45] Given the limitations of available data, no meaningful conclusions could be drawn about loss reductions under the other Argentine concessions.

Figure 3.11 presents the evolution of NRW for 14 large concessions and divestures in Brazil (Manaus, Tocantins, Campo Grande, Campos, Limeira, Paranagua, Petropolis, Itapemirim, and Prolagos), Bolivia (La Paz–El Alto),

44. The largest lease contract in South Africa is for the city of Queenstown, serving a population of 180,000 people in 2007. The lease in Stutterheim and the concession in Dolphin Coast were for smaller systems serving fewer than 50,000 people each. No data could be found in existing publications on the evolution of NRW in the Nelspruit concession (270,000 people, active since 1999) or the Fort Beaufort lease (1995–2000).

45. Between 1992 and 1998, the number of connections increased by 20 percent while water production went up by only 4 percent (Alcazar, Abdala, and Shirley 2000), which is unlikely to have happened without significant reduction in physical losses. And for the period 1999–2003, Casarín, Delfino, and Delfino (2007) report that, overall, leaks went down from 1.45 million cubic meters per day to 1.23 million cubic meters.

Box 3.3

**Introducing Special Incentives for Efficiency in Affermages in Western Africa**

The affermage in Guinea (1989) was the first PPP to be awarded in a developing country in decades. Its performance in reducing NRW and improving the bill collection ratio was unsatisfactory. One key lesson from its implementation was that the incentives for operational efficiency that are contained in a standard affermage contract were insufficient to foster good operational performance from the private operator.

Building on this lesson, designers of the Senegal affermage, which was awarded in 1996, included specific contractual targets for NRW reduction as well as bill collection, backed by financial penalties for noncompliance. The operator's remuneration is not simply based on a fee per cubic meter multiplied by the volume of water actually sold and collected; it is determined by a notional sales volume based on the amount of water actually produced, factored by predetermined annual targets for NRW and collection rates. Whenever the operator fell short of the NRW and bill collection targets, the notional sales volume would be lower than the actual sales, penalizing the operator. This adaptation of the affermage scheme introduced specific contractual targets that are, in fact, more typical of a management contract.

Another innovation in Senegal's PPP was that the private operator was made responsible for directly carrying out a portion of the network's rehabilitation. This included replacing 17 kilometers of pipe, 14,000 water meters, and 6,000 connections each year, to be financed by the operator through cash flow from its operating fee. This approach provided the operator with more flexibility to identify and rapidly carry out actions to reduce water losses, reducing its dependency on the public asset-holding company.

These adaptations proved effective in bolstering the incentives for the private operator to control and reduce water losses. Senegal has now achieved a level of NRW comparable to that of the best water utilities in Western Europe. This approach was broadly replicated in the Niger affermage contract, which started in 2001, as well as in the affermage for the national water utility in Cameroon that was awarded in 2007. Both use the same incentive formula for calculating the operator's remuneration. The Niger affermage also made the operator responsible for carrying out the replacement of 64 kilometers of pipes, to be financed directly out of its revenues during the first five years.

*Source:* Fall and others 2009.

**Figure 3.11 Evolution of Water Losses under 14 PPPs, by NRW Level in Latin America**

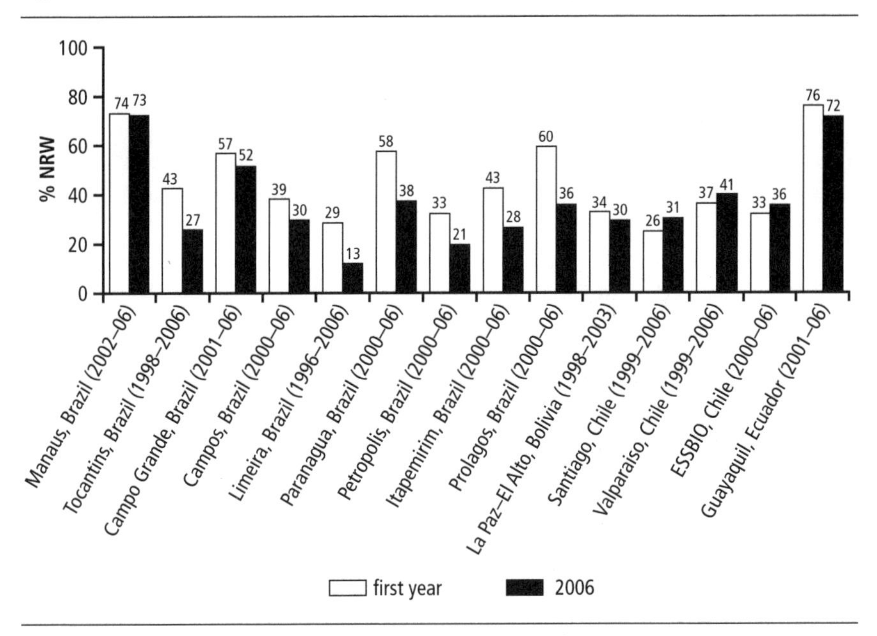


*Source:* National regulators (see appendix A).
*Note:* Not all cases have data available starting from year of takeover by the private operator.
ESSBIO (Empresa de Servicios Sanitarios del Bío Bío) is a water utility in Chile.

Chile (Santiago, Valparaiso, and ESSBIO), and Ecuador (Guayaquil). This sample of PPP projects represents a combined population served of 17 million people. In Brazil, with the notable exception of Manaus,[46] most large PPPs significantly reduced NRW levels. Limeira even stands out for achieving an NRW level of only 13 percent, comparable to the best utilities in developed countries. NRW also went down in La Paz–El Alto and Guayaquil, but the improvement was only marginal.

Chile presents a peculiar situation. Water losses increased after the transfer to private operators, with NRW going up at the national level from 29 percent to 34 percent between 1999 and 2006. In Santiago (5.5 million people), which has about 40 percent of Chile's urban population, NRW went up from 26 percent to 31 percent in that period. Several other utilities also saw their NRW rise by several percentage points. This is paradoxical, because it is widely considered that most Chilean private utilities are

46. The metering ratio in Manaus stood at only 61 percent in 2006, so the NRW figure is not quite representative of actual losses.

well-managed (Bitran and Arellano 2005). Part of the reason for the rise in the NRW percentage could be the expansion of the system—something suggested by the fact that, when measured using alternative indicators, the increase in water losses is rather small.[47] Another explanation may lie in the concept of "optimal level of leakage" (box 3.4). Reducing NRW is not a goal in itself, and it is possible that Chilean public utilities, which had achieved major improvements in operational indicators during the 1990s, may have reduced losses to below the optimal economic level. Under an efficient regulatory framework, a profit-seeking operator should be expected to economically optimize the level of leaks, and this can mean sometimes letting them go up if the cost of investing in further leak reduction activities exceeds the associated financial benefits (Ducci and Medel 2007).

**Reductions in Water Losses in Asia**

The performance of Asia's seven largest and oldest concessions in reducing NRW levels is shown in figure 3.12. The performance of concessionaires in Asia appears quite diverse, illustrating the fact that to succeed in reducing water losses when starting with a highly deteriorated network, both technical expertise and access to sufficient financing for rehabilitation investment are called for.

The concession in Macao stands out as a very good performer, with NRW at only 12 percent, while the case of the concessions in Manila (the Philippines) and Jakarta (Indonesia) illustrates that, in a badly deteriorated network, an operator cannot significantly reduce water losses without major investments in rehabilitation. The two concessions in Jakarta had a series of difficulties that limited their access to financing, and water losses have remained high after almost a decade of private operation. In Western Manila, the concession was never financially healthy, sliding gradually into bankruptcy until it was terminated and awarded to a different private consortium, and water losses remained high after a decade. In contrast, the concessionaire in the Eastern zone significantly reduced its NRW level, but progress started only after the 2002 tariff-rate rebasing that restored financial equilibrium. The resulting increase in revenues then enabled the concessionaire to undertake a major program of leak reduction and thereby reduce NRW from 51 percent to 30 percent in just three years (Navarro 2007).

47. Nationwide, the number of customers in Chile went up from 3.3 million to 4.0 million between 1999 and 2006, while the network expanded from 30,000 kilometers to 36,000 kilometers. Water losses per kilometer increased from 34 cubic meters per day to 38 cubic meters, and from 300 liters per customer per day to 330 liters. When the 1998 data are used as reference instead of that of 1999, water losses appear stable. In at least the case of the utility ESSBIO, water losses decreased in 2000–06 when measured by these two indicators.

**Box 3.4**

**The Concept of the Economically Optimal Level of Leakage: Illustration from Chile**

Water leaks in a large and complex distribution network cannot be totally eliminated. For each network, there is an optimal level of physical losses, which corresponds to the point at which the incremental financial benefit from further loss reductions is lower than the corresponding cost. This optimal level varies considerably among utilities, depending on their specific situation and cost structure (particularly production costs and whether the network is gravity-fed or uses reboosting stations). In a long-term concession in which the private operator has control over investment and starts with a high level of leaks, it is usually profitable to invest in leak reduction. But many public utilities have achieved good technical results in reducing leaks. In such a case, when the utility is transferred to a concessionaire, the issue becomes not just technical but also economic, because a concessionaire is usually much more motivated than a public operator by financial incentives. Even though the concessionaire might have the expertise to reduce losses further, investing in leak reduction activities might just not be the most economical option, depending on each situation.

This may be the situation in Chile, whose overall level of NRW increased from 29 percent to 34 percent nationwide between 1999 and 2006, even though there is widespread agreement that the water services have been operated efficiently under the supervision of a capable regulator. Because the regulator's methodology for setting tariffs relies on a model company that has a 15 percent level of NRW (lower than that achieved by most utilities), these increases in NRW have not been passed on to customers through higher tariffs and have, therefore, directly affected the operational margin of the operators (Ducci and Medel 2007). Faced with such incentives, several private companies may have decided to let leaks increase until they reached the economically optimal level the companies had calculated.

### Performance of Management Contracts In Reducing Water Losses

For 14 management contracts, figure 3.13 compares the NRW level at the entry of the private operator with the NRW level achieved at the end of the contract. These projects represent a combined population served of close to 15 million people (about 75 percent of the population in developing and transition countries who have been served for at least three years by private operators under management contracts).

**Figure 3.12 Water Losses under Seven PPPs, by NRW Level in Southeast Asia**

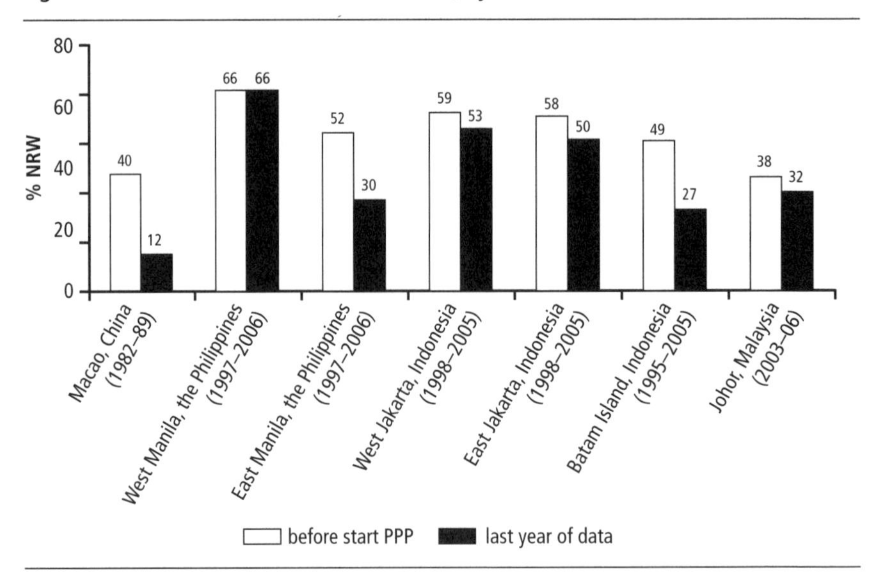


*Source:* Regulator or company data.

**Figure 3.13 Water Losses under 14 Management Contracts, by NRW Level**

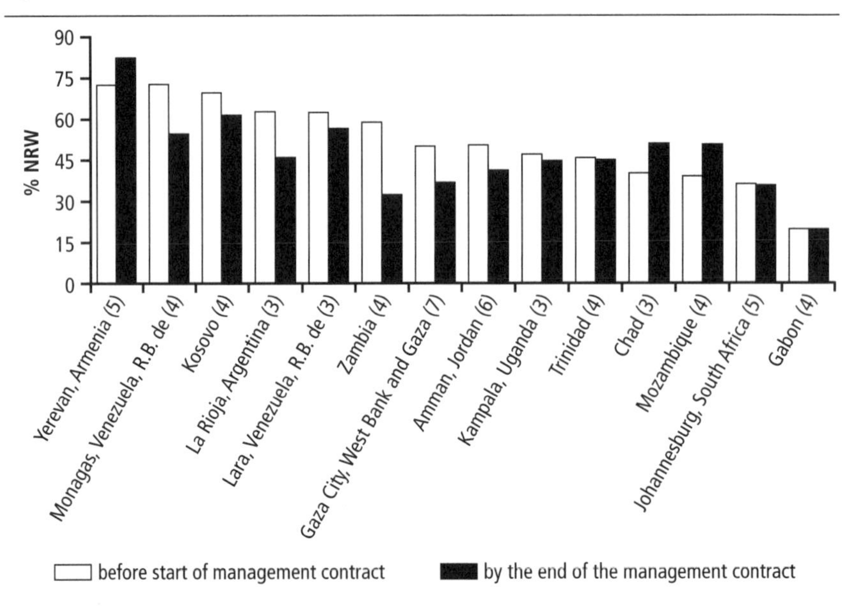


*Source:* Author's calculations based on various sources (see appendix A).

*Notes:* Years of operation are indicated in parenthesis.

The overall performance of management contracts in reducing NRW levels has been mixed at best. Among the 14 contracts for which data were gathered, fewer than half achieved a sizable reduction (Gaza City in West Bank and Gaza, Kosovo, Zambia, La Rioja province in Argentina, Monagas state in República Bolivariana de Venezuela, and Amman in Jordan). No significant change occurred in five other cases (Gabon, Trinidad, Lara state in República Bolivariana de Venezuela, Johannesburg in South Africa, and Kampala in Uganda), and in three cases the NRW level even deteriorated (Chad, Mozambique, and Yerevan in Armenia).

This lackluster result is not entirely surprising, considering the short duration and inherent limitations of management contracts. Such arrangements are better adapted to dealing with commercial losses (which have a rapid payback and require little investment), than with situations where most of the losses come from physical leaks (which usually require large investments and pipe repairs over many years). The diversity of outcomes probably reflects the diversity of the situations encountered, as well as the specific design of each contract.

An important element to consider is the correlation between water losses and service continuity. In a system in which water distribution has been subject to rationing but in which the average number of hours of service is gradually being increased, the average pressure in the network also tends to go up. The higher pressure generates new pipe breaks and more water losses unless significant rehabilitation and improvement in network hydraulics take place simultaneously. This situation has practical consequences for evaluating the performance of an operator: in situations where service continuity is being improved, the evolution of the NRW indicator does not fully capture the actual effort for improving the network's hydraulic functioning and for controlling leaks.[48] In that regard, it is significant that many of the management contracts reviewed in the study were in such situations, starting under conditions of water rationing and achieving notable improvements in service continuity. This makes interpreting NRW data dificult.

A detailed analysis of the situation of each project would go beyond the scope of this study, but a few cases illustrate the complexity of analyzing the evolution of NRW in specific projects.

48. For example, in a situation of intermittent service, an operator can easily reduce the NRW level by just reducing the number of service hours, which also directly reduces the average pipe pressure and, hence, the amount of leakages. Because improving the number of service hours also increases the average pressure in the distribution network, even maintaining NRW levels under such conditions supposes that the operator takes active actions to control leakages and to improve the network hydraulic.

In **Johannesburg** (South Africa), the responsibility for NRW reduction was shared between the private operator and the city under a complex framework. No progress was made in reducing NRW levels under the management contract, but a closer look at the situation shows that physical leaks in most of the network stood at only 15 percent, and the largest source of losses came from huge wastage by customers in the townships (whose consumption was not metered). The partners started to address this sensitive issue only in the last two years of the contract, which ended before sizable results could be felt (Marin, Mas, and Palmer 2009).

The management contract in **Yerevan** (Armenia) focused on switching from estimated to metered billing for residential customers and on reestablishing continuous service—something it largely achieved. The NRW level went up significantly, as a result of both the increase in average network pressure and a 150 percent drop in billed volumes. But reducing water losses was not a priority objective: the management contract focused first on reestablishing continuous service.

The case of **Amman** (Jordan) is particularly important because one of the priorities of this management contract was to reduce water losses. The utility faced acute water scarcity, and the contract was carried out in parallel with a major network rehabilitation program financed by donors. The execution of the investment program, which was essential for reducing water losses, was the responsibility of a government agency. The implementation of the management contract proved complex, with repeated delays in completing civil works and several contractual adjustments. It offers important lessons regarding the need for smooth coordination between the partners for the implementation of civil works, as well as for realistic contractual targets (box 3.5).

### *Improvements in Bill Collection*

It is widely accepted that private operators are usually efficient in collecting bills, for the simple reason that they are profit oriented and collecting bills directly affects their financial results. The rate of bill collection can be improved either by enforcing stricter collection policies or by improving service quality, which, in turn, increases customers' willingness to pay their bills. In reviewing trends in bill collection, however, one often finds it difficult to tell which of these elements has played a greater role.

How much a private operator can increase the rate of bill collection obviously depends on the starting level, but it is also influenced by cultural and country-specific issues. In Senegal, for instance, the population had a strong tradition of paying their water bills, and the collection ratio was already very good when the private operators took over. Elsewhere, private operators often have to face well-entrenched habits of nonpayment of water bills

Box 3.5

**Reducing Water Losses by Combining a Management Contract with Large Rehabilitation Work in Amman, Jordan**

The management contract in Amman illustrates the difficulty of assessing the contribution of the private operator in a management contract. The PPP was just one element of a major investment project (about US$200 million) to completely rehabilitate Amman's water distribution network. The plan was to move from a poorly designed hydraulic system to one with well-delineated zones that are gravity fed through reservoirs. The management contract was supposed to ensure that an experienced operator would be in place so that this major structural change could be handled smoothly, reducing service disruptions and maximizing operational benefits from the new infrastructure.

There was acute water rationing in the city (customers received water for less than four hours per day on average), and reducing water losses was a top priority of the program, but it depended both on the rehabilitation program implemented by the government and on operational improvements to be made by the private operator. These dual responsibilities were not clearly acknowledged in the original drafting of the contract. In addition, ambitious targets were imposed on the operator, backed by swift financial penalties: the NRW level was to be reduced by 10 percentage points in the first year of operation, down to 25 percent water loss (broadly halving the NRW level) by the end of year four.

Difficulties started early on, as the government agency in charge of implementing the civil works experienced major contract delays. Further delays arose during execution as a result of the complexity of coordinating the work of many contractors. After tense discussions in the first two years, it was recognized that the private operator could not be held liable for failing to meet the contractual NRW targets and that the targets had been overestimated. A special project monitoring unit was also set up to help the government better play its role as counterpart in the partnership.

As the program proceeded, another problem occurred: with the gradual reduction in water rationing, the increase in average network pressure caused a jump in the number of water leaks. The operator had to repair as many as 55,000 leaks in 2004. By the end of the contract, it had replaced on its own about 600 kilometers of pipe, or close to 10 percent of the whole network. The management contract was extended twice in order to

*(continued)*

**Box 3.5**

**Reducing Water Losses by Combining a Management Contract with Large Rehabilitation Work in Amman, Jordan *(continued)***

keep the private operator in place until the end of the capital expenditure program in 2006. By the end of the contract, NRW had been reduced from 51 percent to 42 percent. This was a notable improvement, even though it fell far below the original target. At the same time, the average number of hours of service was doubled.

Important lessons can be learned from this case. Though the parties finally agreed that the original NRW target of 25 percent was unrealistic, their protracted negotiations on this subject distracted them in the early years from focusing on more productive tasks. Tracking the operator's performance was made difficult by its dependency on the government's timely execution of the investment program. Finally, the hydraulics of the network were being profoundly altered, so that the basic reference point for measuring leaks was, in fact, constantly changing. In such a context, using the NRW percentage as the sole contractual indicator of the operator's performance to track water losses and impose stiff financial penalties appears, in retrospect, to have been a mistake.

*Source:* El-Nasser 2007.

or a variety of legal impediments that prevent them from enforcing payment. This can make progress difficult. A culture of nonpayment usually develops in reaction to many years of poor service, and achieving a behavioral change in the population takes time.

### Improvements in Collection Rates in Latin America

Latin America presents the largest sample of PPPs with available data on bill collections, giving the opportunity to observe differences among several countries. Figure 3.14 shows the increase in bill collection ratio achieved after a few years of private operation in 16 large concessions, leases, and divestitures (representing a combined served population of 28 million).

Performance has been quite diverse across countries. Significant progress was made in just a few years by several PPP projects in **Brazil** (Campo Grande, Campos, Limeira, Niteroi, Manaus, and Tocantins) and **Colombia** (Barranquilla, Cartagena, and Monteria), as well as in **La Paz–El Alto** (Bolivia) and **Guayaquil** (Ecuador). In **Chile,** most regional public utilities had already achieved high rates of bill collection, and collection ratios have usually remained at above 97 percent since the transfer (Valparaiso

**Figure 3.14 Increases in Bill Collection Ratio under PPPs in Latin America**

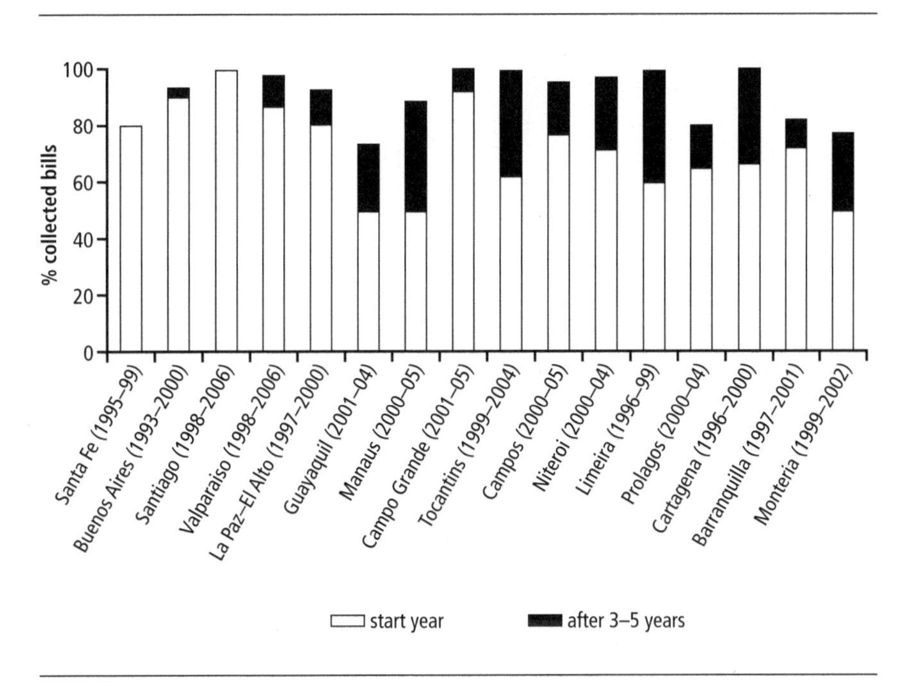


*Source:* Author's calculation based on public sources (appendix A).

was an outlier, allowing the private operator to make sizable gains). In **Argentina**, the performance of concessionaires in improving the collection ratio has been rather disappointing. Ducci (2007) indicates that only moderate improvement was achieved in Buenos Aires (up from 90 percent to 94 percent in seven years) and none in Santa Fe (still at only 80 percent by 2001). The concessionaire in the Argentine provinces of Salta, La Rioja, and Corrientes also encountered difficulties, as it was not allowed to cut off service in cases of nonpayment (Yepes 2007).

It is notable that many PPP projects that achieved significant gains still had collection rates below 90 percent or even 80 percent after several years of private operation. This is the case especially in Colombia, even though in the three cases mentioned above the private operators have performed very well on other dimensions. This illustrates how long it can take to change a culture of nonpayment. In the successful case of Cartagena, the operator achieved a satisfactory ratio of 95 percent only after a decade of sustained service improvements.

**Bill Collection from Various Customer Categories in Sub-Saharan Africa**
The performance of private operators in Western Africa for collecting bills presents a special picture. The collection of bills from residential customers has often far outpaced the overall collection rate, due to frequent difficulties in collecting bills from government agencies.

Several private operators have had good success in the region in collecting bills from domestic customers. In **Senegal**, the level of bill payment by residential customers was already at 98 percent under public management, and this was maintained under private management. In **Niger**, the collection rate from households improved from 93 percent at the start of the affermage contract in 2001 to 97 percent in 2006. Good collection rates were also achieved from residential customers in **Côte d'Ivoire** and **Gabon**. A few PPP projects performed poorly: no progress was achieved in **Guinea** over a decade of private operation, with the collection ratio for residential customers remaining at about 60 percent; and **Dar es Salaam** (Tanzania) is a rare case in which the residential bill collection rate went down after the private operator took over (that contract was terminated in its second year).

The high collection rates achieved in several cases are probably linked to the fact that in many of these countries only the richer portion of the urban population has access to piped water through a household connection; the poorest families use community standpipes or purchase water from neighbors and, therefore, do not get monthly water bills. This is well-illustrated by the experience in Senegal, which has the highest coverage through household connections in the region (79 percent in 2006) thanks to the implementation of a large subsidized connection program. Senegal's high collection rate from residential customers conceals the fact that the rate of disconnection has been in the range of 10 to 15 percent in recent years—meaning that, in practice, a sizable proportion of the poor households who benefited from the subsidized connection program ended up being disconnected at some time (most of these households still access piped water by purchasing it from neighbors). A similar phenomenon has been observed in Côte d'Ivoire.

A major issue in all water PPP projects in Sub-Saharan Africa has been the serious difficulty of collecting bills from public buildings and government agencies, which typically represent a sizable portion of the revenues of water utilities in the region. Erratic payment by public customers has been a recurrent problem, even in an erstwhile successful PPP such as Senegal's. Private operators are ill equipped to collect bills from accounts that ultimately belong to their contractual partners, and donors regularly have had to step in and remind governments of their contractual obligations. Special mechanisms have been developed gradually to mitigate the problem. In Senegal, the operator can now appeal directly to the Ministry of Finance, which directly intervenes in the case of nonpayment by a public agency. In Niger, a system of advance payments has been put in place, with estimated monthly bills for all public agencies paid every month by the Finance Ministry, subject to adjustment at year's end.

**South Africa** presents a special situation. During the apartheid years, nonpayment of utility bills by township residents was widely viewed as an act of civil resistance and became the norm. This behavior has continued since the end of apartheid; it affects all water utilities and has proved very hard to change. Because of social sensitivities, lease contracts (such as those in Queenstown and Stutterheim) were designed with the peculiarity of leaving the responsibility for bill collection to the municipal government. As for concessions, the largest one in Nelspruit experienced major difficulties in collecting bills, with the collection ratio going down to less than 30 percent, while the one in Dolphin Coast was more successful, going up from 75 percent to 97 percent in four years (Palmer 2003).

### Improvements in Bill Collection under Most Management Projects

Most management contracts have performed well in improving bill collection. Although the bills received by customers are still issued by a publicly owned utility, the sharing of responsibility between the government and private operator for bill collection has varied among contracts. For the 15 management contracts for which data were available, figure 3.15 compares the bill collection ratio before the entry of the private operator with the one at the end of the contract.

Bill collection seems to be the dimension of performance in which management contracts have been most consistently effective. All documented management contracts have resulted in sizable improvements in bill collection in a relatively short time. These improvements were often accompanied by improvements in service quality (continuity of supply and/or customer service), as happened in Albania, Gaza City (West Bank and Gaza), Zambia, Amman (Jordan), Kampala (Uganda), La Rioja (Argentina), and Yerevan (Armenia). But there are also cases such as Guyana and Trinidad where the collection ratio was improved despite little objective improvement in service quality, suggesting that the main reason there was stricter enforcement.

In **Gaza City** and **Amman**, notable improvements were achieved through a combination of stricter enforcement, extensive education campaigns, close collaboration between the operator and public authorities, and sizable gradual improvements in service quality. In Amman, the collection rate started at 83 percent. Improvement in the first two years came from stricter collection policies, but the rate still kept rising afterward as progress was made on reducing water rationing and improving customer service. After seven

**Figure 3.15 Improvements in the Bill Collection Ratio under 15 Management Contracts**

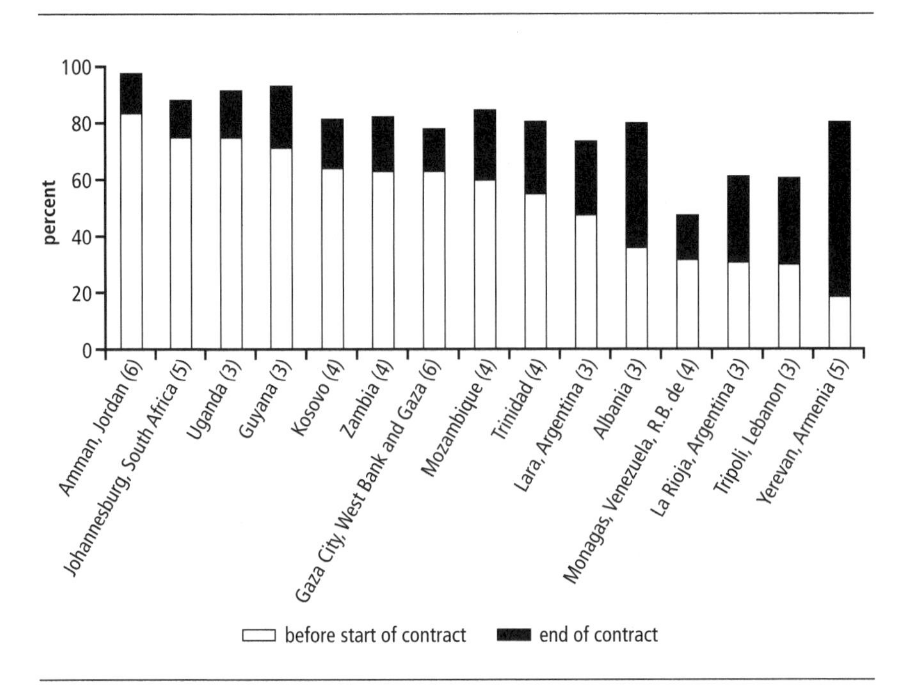


*Source:* Author's calculations based on various sources (appendix A).

*Note:* Years of operation are indicated in parenthesis. For Gaza City, the data are for the period up to 2000.

years of private operation, the contract ended in 2006 with a collection rate at 97 percent (El-Nasser 2007). In **Gaza City**, bill collection went up from 63 percent in 1996 to 78 percent in 2000. However, the sharp degradation of general conditions in Gaza City after 2001 made bill collection policies difficult to enforce. The management contract ended with a collection ratio at 53 percent, a level even lower than when it took over the utility (Jme'an and Al-Jamal 2004).

The most spectacular improvement was achieved in **Yerevan,** with the collection rate going up from less than 20 percent to 80 percent in five years. This was achieved through close collaboration between the operator and the government, and in parallel with sizable improvements in service quality (box 3.6).

**Box 3.6**

**The Management Contract: Remarkable Success in Improving Bill Collection in Yerevan, Armenia**

The case of the management contract in Yerevan illustrates the link between improvement in service quality and the collection ratio, as well as the need for government to support the private operator with proper policy initiatives. The collection rate started at a very low level of 19 percent. Initial efforts to improve bill payment resulted in some improvement, but by 2002, more than half of the billed revenues still went uncollected.

Changes in customer behavior and more improvements in the rate of payment did not begin until the government issued a decree (in 2002) that allowed the operator to disconnect nonpaying customers and passed a law that included provisions for partial forgiveness of customer debts in return for individual meter installation. Many households then accepted the installation of meters and negotiated a partial repayment of bills in arrears. This boosted the collection ratio in 2003, which was sustained as service continuity was gradually restored in most parts of the city. All this was complemented by education campaigns and specific actions to repair leaks in apartment buildings. The result of all these actions was remarkable: the collection rate stood at about 80 percent at the end of the management contract, a fourfold increase over the pre-PPP level, reflecting a major behavioral change in the population and a real shift in perceptions of the value of water supply and sanitation services.

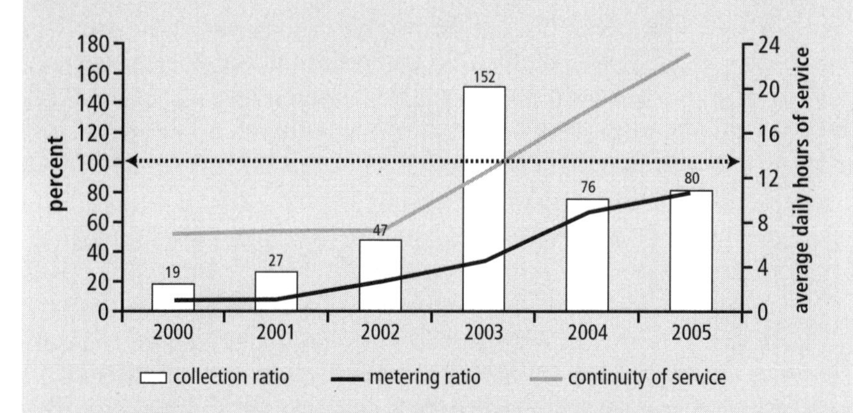


*Sources:* Mugabi and Marin 2008; World Bank project reports.

### *Productivity and Labor Issues*

Workforce issues are important elements of water PPP projects, and falling employment levels have been a key concern of opponents of private participation in water utilities. Findings from Andrés, Guasch, and others (2008) and Gassner, Popov, and Pushak (2008a) confirm that introducing a private water operator usually results in lower staffing levels and higher labor productivity. This raises a dilemma: workers are assets, and governments and donors cannot ignore social issues linked to downsizing. But labor costs are also a major component of utility costs, so productivity cannot be ignored either, regardless of whether a utility is privately or publicly managed. Furthermore, moving from public to private management often entails a major change in corporate culture, affecting staffing and salary levels, job qualifications, work rules, and promotion practices. For the provision of better and more efficient services, it is often necessary to replace staff members who have limited or no qualifications with others who are more qualified, making some redundancies unavoidable.

#### Private Operators, Labor Productivity, and Employment Levels

The evolution of the labor productivity ratio—calculated as the number of staff members per thousand customers—was examined for 17 large long-term PPPs, starting from the time the private operator took over (figure 3.16). Most private operators have significantly improved the labor productivity ratio, usually by combining a reduction in the number of staff with an expansion in the customer base through coverage expansion. Not surprisingly, smaller gains were achieved where the productivity ratio was initially better, such as in Chile, Barranquilla (Colombia), Guayaquil (Ecuador), and Salta (Argentina).

In several cases, preparing for the transition to a water PPP involved layoffs carried out beforehand by the government, in an attempt to make the partnerships more attractive to private operators. This was especially so in the most obvious cases of overstaffing—an issue the private sector was reluctant to deal with directly because of the social sensitivity involved. Such layoffs are not reflected in figure 3.16.

To present a fuller picture of what happened as a result of a PPP process, figure 3.17 shows the total layoffs that took place for 10 large PPP projects in Latin America. The figure combines the layoffs initiated by the contracting government in the 12 months before the award of the contract with those that were carried out directly by the private operator in the year (or sometimes two years) after the takeover. In at least a few documented cases, the unions negotiated generous packages (Chile, Buenos Aires [Argentina], and Guayaquil [Ecuador]).

**Figure 3.16 Evolution of the Labor Productivity Ratio for 17 Large PPPs**

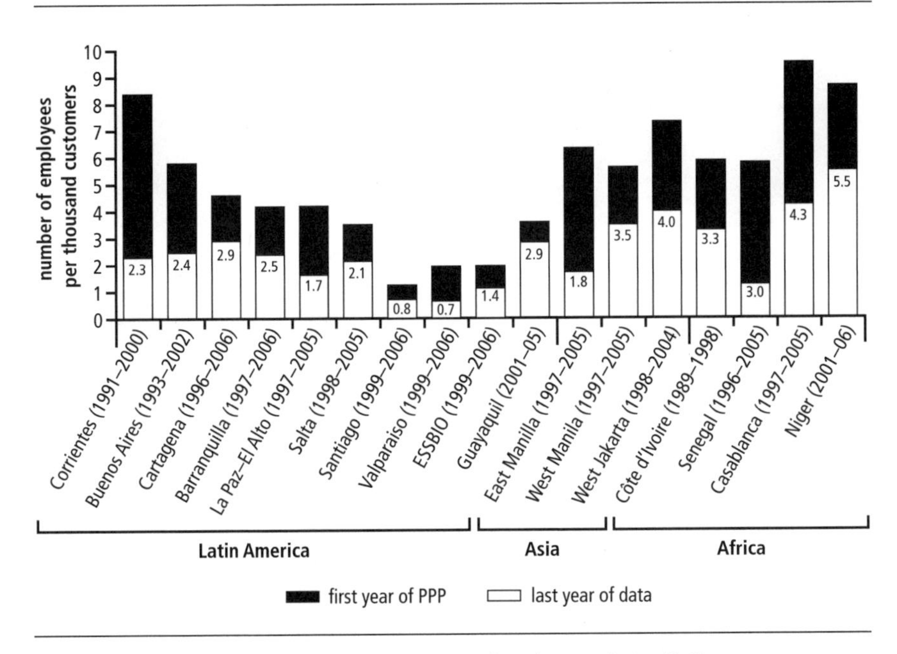


*Source:* Author's calculation based on public data or consultants' reports (appendix A).

*Notes:* The figure compares the ratio in the first and last years of available data, or year 10 for PPPs that have been in place for a longer period. The labor productivity ratio for year one corresponds to when the private operator took over. In several cases important layoffs had already taken place before the signing of the contract (as in Buenos Aires, Guayaquil, and Manila). The Casablanca figure includes staff for both water and electricity.

ESSBIO (Empresa de Servicios Sanitarios del Bío Bío) is a water utility in Chile.

Figure 3.17 confirms that several water PPP projects have been accompanied by massive initial layoffs of utility staff. In the documented sample, the layoffs ranged from about minus 25 percent in Salta (Argentina), Barranquilla (Colombia), and Santiago (Chile) to as much as minus 65 percent in Cartagena (Colombia).

These layoffs were, in large measure, justified by overstaffing, which came as a result of years of political interference and clientelism. The variations in the magnitude of downsizing can be largely explained by how much excess staffing existed before the reform. In Cartagena, where labor productivity stood at the dismal ratio of 15 employees per 1,000 customers in 1994, two-thirds of the staff were made redundant. In Buenos Aires, the state-owned utility OSN (Obras Sanitarias de la Nación) had about 8,000 employ-

**Figure 3.17 Employment Reduction Associated with Implementation of 10 Large PPPs in Latin America**

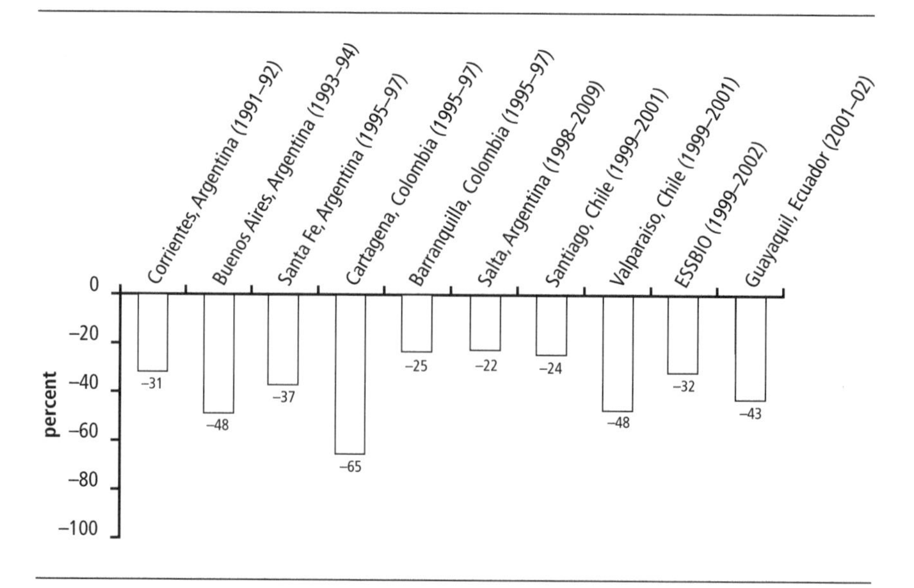


*Source:* Author's calculation based on various sources.

*Note:* ESSBIO (Empresa de Servicios Sanitarios del Bío Bío) is a water utility in Chile.

ees, equivalent to a labor productivity ratio of 9 per 1,000 customers, and work practices were notoriously poor (Idelovitch and Ringskog 1995). Barranquilla had a different situation, because a significant staff reduction had already been carried out gradually over the previous five years under public management. The labor productivity ratio stood at a more reasonable 5.5 employees per 1,000 customers in 1996, so the downsizing that was needed during the PPP process was more modest. However, in the case of Chile, significant layoffs still took place even though the labor productivity was already satisfactory when the private operators took over.

It is noteworthy that after the initial staffing adjustments were made, the number of employees of the documented PPP projects usually rose or remained stable. This result occurred partly because suitably qualified personnel had to be recruited after the initial layoffs, and partly because many PPP projects rapidly increased access to previously unserved customers. With larger customer bases, further productivity gains could be made without having to resort to further layoffs. In Buenos Aires, after the sharp initial decrease, total staffing actually increased by 450 employees during the first four years. The same situation was observed in Guayaquil, associated with the fast expansion of coverage there (the number of water connections went up by 65 percent in the first five years).

In practice, downsizing has been concentrated mostly in Latin America, although it also occurred in Manila (the Philippines), where staffing levels were reduced by about 40 percent. Elsewhere, a number of projects dealt gradually with excess staffing, through natural attrition. This occurred in several PPP projects in Sub-Saharan Africa. In Senegal and Gabon, modest reductions in staffing of only 15 percent and 10 percent, respectively, took place over a decade, and in Niger, staffing levels have thus far been unaffected. Reductions were modest, too, in Maputo, Casablanca, and Jakarta, with staff numbers falling by less than 20 percent after several years.

The issue of subcontracting deserves special notice. Private operators tend to use subcontractors whenever possible because these provide more flexibility. However, subcontracting makes it harder to assess the actual impact of PPP projects on net employment, because it can partially offset job losses within the utility. In Chile, for instance, the recourse to subcontractors increased with the transfer to private operators. The national regulator recently started to collect data on subcontracting by the water utilities. It found that, in 2006, Chilean private utilities had more workers employed through subcontractors than through direct employment. For instance, the private utility in Santiago (Empresa Metropolitana de Obras Sanitarias, or EMOS) had about 1,100 direct employees but also employed an equivalent of an additional 1,500 workers through subcontractors. Although no data on subcontracting exist for the period when the utilities where transferred to the private sector, it is not unlikely that the staffing reductions in Chile (from 6,600 in 1998 to 4,500 by 2006 for the combined regional utilities) were at least partially offset by increases in employment by subcontractors (Ducci and Medel 2007).

### Impact of PPPs on Salaries and Working Conditions

The impact of PPP projects on labor goes beyond labor productivity and staffing levels. Water utilities are in the business of providing a service, so although staff are indeed a cost, they also are an essential asset. Providing good service to customers cannot realistically be done with a dissatisfied workforce.

The impact of private operators on labor costs—looking at the evolution of salary costs instead of just employment levels and productivity ratios—has not been analyzed in published studies. This study collected only a small amount of data on the evolution of average salaries. Circumstantial data on Niger, Casablanca (Morocco), and Manila (the Philippines) suggest a possible upward trend under private management, but in Côte d'Ivoire, average salaries in the PPP went down in real terms by 25 percent in 15 years.[49] This important issue deserves further study.

### Labor Aspects of Management Contracts

In management contracts, unlike other types of PPPs, the utility's staff members retain their status as civil servants; although the private operator is carrying out the day-to-day management, it typically has little or no authority over staffing levels, hiring and firing, salaries, or promotions, except in its advisory role. However, because management contracts are usually implemented as part of a broader reform to improve sector performance, operators have often had to deal with overstaffing.

The evolution of staffing levels in water utilities under management contracts has been extremely diverse. In some cases, staffing levels were reduced through voluntary reassignment to other government or municipal departments, such as in Antalya (Turkey; down from 500 to 200 employees), and Amman (Jordan; down from 1,600 to 1,400 employees). In other cases, the number of employees remained broadly stable, such as in Johannesburg (South Africa), Kampala (Uganda), and Lara and Monagas (República Bolivariana de Venezuela). In yet others, the number of employees went up significantly, as in Yerevan (Armenia) and Zambia.

Private operators' biggest contribution to human resources under a management contract lies in the transfer of expertise and change in corporate culture. This contribution is very difficult to capture in numbers but plays a major role in achieving sustainable improvements in service quality and operational efficiency. This issue has rarely been addressed in the published literature but was documented in the case study of the Johannesburg management contract (Marin, Mas, and Palmer 2009), prepared as part of this review (box 3.7).

### *Conclusions on PPPs and Operational Efficiency*

The previous discussion cites a large number of cases in which private operators have been effective in improving the efficiency of water utilities in three key dimensions: water loss reduction, bill collection, and labor productivity. This study's indirect and incomplete approach to measuring efficiency still makes it possible to draw some relevant conclusions from the literature and the case studies reviewed.

---

49. In the Niger affermage, the private operator gave a 20 percent salary raise in the first year. In Casablanca, the operator reported that the salary received by a fieldworker is twice the national average for a similar job, and a foreman gets 60 percent more. In Manila, the salaries are more than three times higher than those in public utilities (Navarro 2007). In Côte d'Ivoire, average salaries went down by 25 percent while labor cost per cubic meter sold was halved, and efficiency savings were largely passed to customers, because the tariff went down by 35 percent.

**Box 3.7**

**Using a Management Contract to Carry Out a Complete Corporate Reorganization of the Water Utility in Johannesburg, South Africa**

Before 2000, responsibilities for water and sanitation services in Johannesburg were spread across six separate municipal departments: four geographical departments were in charge of water distribution and sewerage networks (answering to four local councils); one department was in charge of the operation of wastewater treatment plants; and the central level of the municipality directly handled all matters related to customer relationships, revenue management, procurement, and finance. This fragmented structure had generated a "silo mentality" among the staff, with a dilution of responsibility and accountability, and customer service was notoriously poor. To remedy this situation, Johannesburg Water was established in 2000 as a new corporatized public utility responsible for water and sanitation services. The main rationale for bringing in a private operator under a five-year management contract was to establish Johannesburg Water as a viable and efficient water utility.

A major part of the job of the private operator was to organize the newly consolidated utility, putting the proper work procedures in place and training employees. The goal was not just to design a new organization chart; it was, most important, to instill a new corporate culture focused on service and efficiency. This was a major challenge with employees who had previously operated under an old-fashioned bureaucratic culture. Implementing such a change was a long and gradual process, in which the daily coaching by the operator staff played a major role. One example of the many measures taken to foster change was empowering line managers to take more initiative in their daily job, as long as this would benefit the quality of service to customers or result in efficiency savings. The average salary per employee went up by 23 percent in real terms during the management contract.

A major effort to promote a new generation of managers and professionals was carried out in parallel with this cultural change. The 693 promotions that occurred during the management contract mostly benefited staff members belonging to groups that previously had faced discrimination in the apartheid era. Although the total number of staff members remained fairly stable (rising from 2,500 in 1999 to 2,600 in 2006), 945 skilled employees were recruited during the management contract, again largely from previously disadvantaged groups.

*Source:* Marin, Mas, and Palmer 2009.

### Operational Efficiency of Concessionaires

Analyzing the operational efficiency of private operators under concessions is made difficult by the interaction between investment and operation. The limited evidence on this subject in the published literature is mixed. In Chile, Bitran and Valenzuela (2003) reported that efficiency improved significantly during the first two years after the entry of private operators. In Argentina, Estache and Trujillo (2003) used a total factor productivity methodology and concluded that efficiency gains had been significant in Buenos Aires and Salta,[50] but Casarín, Delfino, and Delfino (2007) estimated that in Buenos Aires, most of the efficiency gains had come from initial layoffs. Several authors highlighted that in Argentina, the regulatory scheme put in place for concessions was weak and not conducive overall to efficiency gains.[51]

In Manila (the Philippines), a detailed assessment of the operational efficiencies of the two concessions in 1997–2002 was carried out by the regulator with the assistance of foreign experts.[52] The assessment found that the concessionaire in the Eastern zone (Manila Water) had achieved the efficiency level of its initial proposal. But although significant cost savings had been achieved, sales volumes were lower than expected, which affected the financial equilibrium. The outcome was different for the Western zone concession, where the efficiency savings of the initial proposal were not achieved—one of the reasons why it ended up in bankruptcy.[53]

This study did not address the efficiency of concessionaires for investments.[54] As for operational efficiency, the data reviewed suggest that concessionaires have generally had a positive impact on overall efficiency, though with some variations. In Colombia, several concessions (notably Barranquilla and Monteria) achieved significant progress in the three performance indicators under analysis. So did the concession in La Paz–El Alto (Bolivia),

50. These were the only two cases with enough data to make reasonable estimates. The authors indicated that their estimate of 2 percent gains per year on average was a rough one, and probably in the upper bound.

51. See Solanes and Jouravlev (2007) on Buenos Aires and Yepes (2007) on the provincial concessions in Salta, Corrientes, and La Rioja.

52. This assessment was based on a methodology that compared actual operating costs with the financial projections that were submitted in the original bids, adjusted for input price changes and volume changes.

53. A possibly more important factor was that the peso devaluation during the 1996–97 Asian financial crisis sharply raised the cost of servicing the ongoing debt of the previous public utility, which had been mostly transferred to the Western zone concession.

54. In principle, it is expected that the concessionaire would make efficient investment choices, with an impact on operational efficiency, service quality, and access, as well as execute investments in a timely and cost-efficient manner because of financial incentives. But concerns have been expressed over certain practices of granting civil works to companies linked to the concessionaire. Control of investment is a key aspect of efficient regulation.

which was consistently rated by the national regulator Superintendencia de Saneamiento Básico (SISAB) as the best-performing large water utility in the country. The concessionaires in Macao (China) and Limeira (Brazil) achieved efficiency levels comparable to the best-performing water utilities in developed countries. In contrast, in Argentina, the concessionaires' performance in bill collection has been mediocre, and the low rate of customer metering makes it difficult to assess the evolution of water losses. In Guayaquil (Ecuador), the operator significantly improved the rate of bill collection but made no sizeable progress on water losses. And in Manila, the Eastern zone concessionaire performed well in improving efficiency, while the Western zone concessionaire failed to make gains. In a significant number of cases, data were not available for all three indicators.

**Evidence of Efficiency Gains under Lease-Affermage Contracts: Cartagena, Colombia and Senegal**

Detailed data were available on the mixed-ownership company in Cartagena, Colombia, which operates under a lease contract with the municipal government, and the affermage in Senegal. Together, they provide a good illustration of the overall operational efficiency gains that a successful private operator can achieve (see figure 3.18). The analysis uses a simple approach that compares the evolution of the operating ratio (collected revenues in relation to operational costs)[55] to that of the average tariff level. The operating ratio is driven essentially by two factors: the evolution of operational costs and collection rates (which are controlled by the private operator) and the evolution of the average tariff (which is exogenous). Whenever the operating ratio increases faster than the tariff, efficiency gains are taking place.[56]

As described earlier in this chapter, both these PPP projects achieved significant improvements in the three key efficiency indicators under review (level of water losses, bill collection ratio, and labor productivity). Such improvements are reflected in the rise achieved over a decade in the ratio of collected revenues to operational costs, as shown in figure 3.18. In the case of Cartagena, the initial financial situation was much worse than that in Senegal, allowing for larger gains. That the average tariff went down in

55. This definition of *operating ratio* uses the actual amount of collected revenues every year, which is slightly different from the one commonly used in accrual accounting.

56. Although this simple approach clearly illustrates the efficiency gains achieved, it does not work for all situations. The fact that the operating ratio increases faster than the average tariff level in real terms does not mean that there is no efficiency gain, especially where the initial tariff level was far from full cost recovery, and where a large portion of tariffs corresponds to the financing of investment.

Figure 3.18 Efficiency Gains under Leases-Affermages in Cartagena, Colombia and Senegal

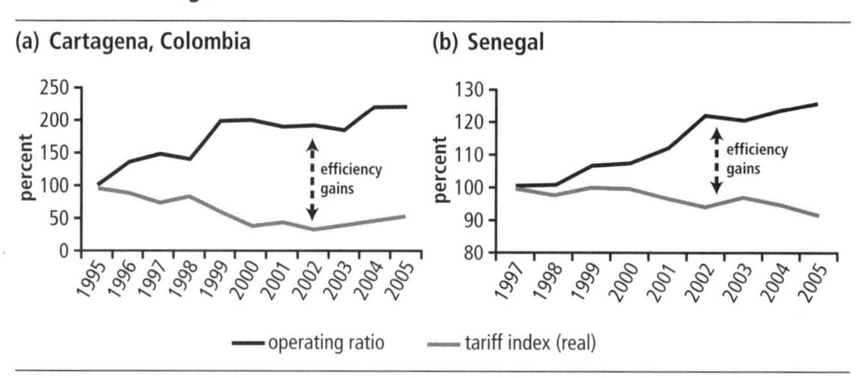


*Source:* Author's calculations.

*Note:* Tariff index represents evolution of average tariff level in real terms (corrected for inflation) with 100 percent representing the level when the PPP project was initiated.

real terms in both cases suggests that a significant portion of the savings was passed to customers.

Data available from other affermage contracts in Sub-Saharan Africa show a mixed picture. The national water utility in Côte d'Ivoire attained a level of operational efficiency comparable to that in the best utilities in developed countries, and its performance proved remarkably resilient in recent years despite civil unrest. In Niger, where the contract follows the same incentive design as that in Senegal, sizable initial gains have been made. In contrast, the overall performance in Guinea proved disappointing despite early improvements, and in Maputo (Mozambique) the many difficulties that have affected the implementation of the partnership, and especially delays in implementing the rehabilitation program, have resulted in little improvement being achieved so far in operational efficiency.

### The Impact of Management Contracts on Operational Efficiency

Under a management contract, the impact of the private operator on operational efficiency is difficult to assess: the private operator typically has only limited control over operational costs, depending on the specific design of each contract. In all the management contracts reviewed, the government retained responsibility for staffing and salary levels.

The evolution of the global efficiency ratio (defined as the ratio of the volume of water billed and for which payment has been collected, divided by the volume of water produced and injected into the network) is shown in figure 3.19. This ratio combines the two indicators—non-revenue water and bill collection—reviewed earlier in this chapter. The figure compares

**Figure 3.19 Global Efficiency Gains under 12 Management Contracts**

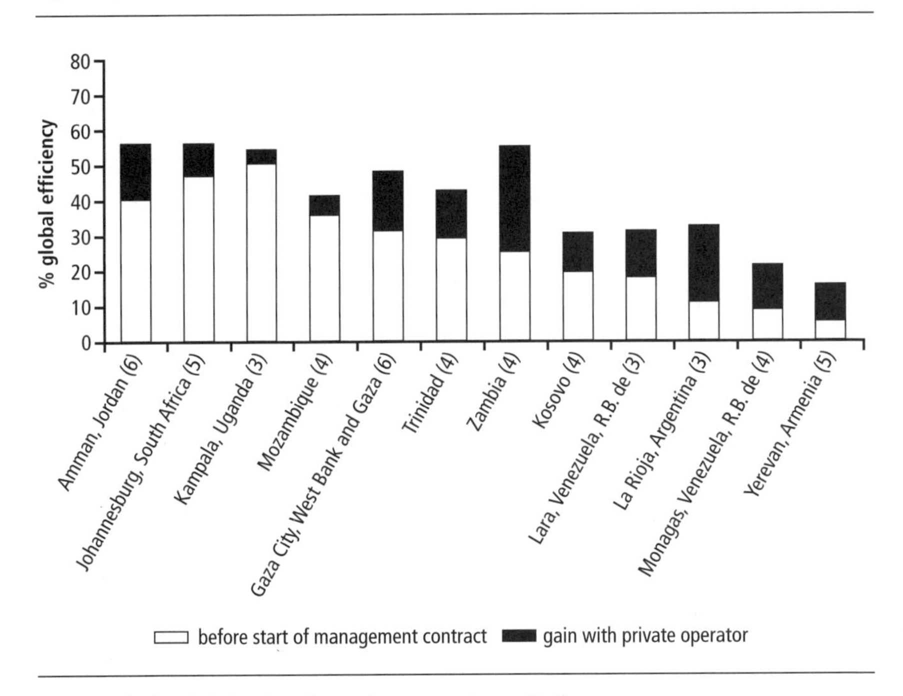


*Source:* Author's calculations based on various sources (appendix A).

*Note:* Efficiency ratio is calculated as the volume of water billed and for which payment has been collected, divided by the volume of water produced and injected into the network.

Years of operation are indicated in parenthesis.

the global efficiency ratio of the water utility before the entry of the private operator and at the end of the contract, analyzing 12 management contracts for which data on both the NRW percentage and the bill collection ratio were available.[57]

A rather consistent pattern appears, despite the many differences in the design of the 12 management contracts. In 10 cases, under private management there was a sizable improvement of the global efficiency ratio, often in the range of 10 to 20 percentage points. Mozambique (Beira and other cities) and Kampala (Uganda) also showed improvement, though of a lesser magnitude. In all cases, these efficiency gains made a notable contribution to improving the financial situation of the utilities concerned.

57. This analysis underestimates the overall gains achieved, because it does not capture other dimensions of improvements in operational efficiency that might have taken place, such as with chemical (treatment) uses and energy consumption.

It is possible to look at the evolution of operational efficiency in management contracts by applying the same methodology as was used with leases-affermages, that is, analyzing the evolution of the operating ratio against the evolution of the average tariff in real terms. Figure 3.20 gives the examples of Amman (Jordan) and Johannesburg (South Africa), for which detailed data were collected as part of this study and where sizable gains were achieved.

In **Amman,** the improvement in the global efficiency ratio was 16 percentage points in five years of private operation, stemming from a combination of a reduction in NRW and an increase in the bill collection rate. Other gains were made in operating costs, and the management contract resulted in a significant financial turnaround: the Greater Amman water utility went from financial losses in 2000 and 2001 to a net profit of 16 percent of revenues in 2005 and of 23 percent in 2007. The management contract was implemented in parallel with a major capital investment program undertaken by the government, so this is a clear case in which success must be credited to the actions of both partners.

In **Johannesburg,** the overall impact of the management contract on operational efficiency and financial viability has been very positive. The utility went from a negative cash flow in 2001 to positive cash flow in 2004, and posted a net profit by the last year of the management contract in 2006. During the last two years of the contract, an independent national panel ranked Johannesburg Water as the best-performing water utility among large cities in South Africa. A detailed financial analysis conducted to assess

**Figure 3.20 Example of Efficiency Gains under Management Contracts in Amman, Jordan and Johannesburg, South Africa**

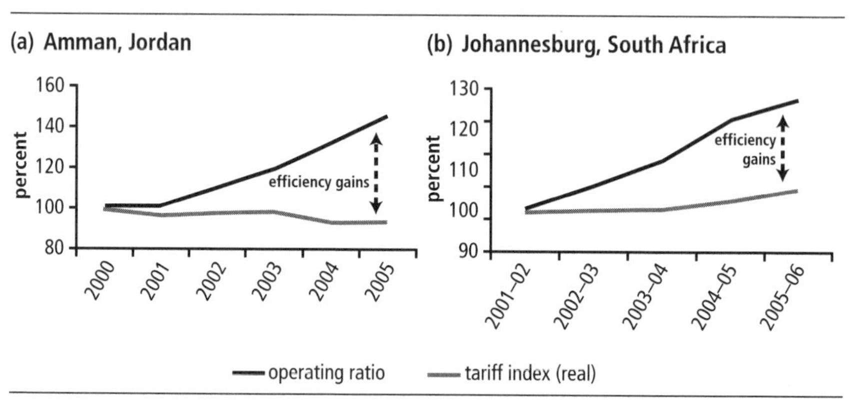


*Source:* Author's calculations.

*Note:* Tariff index represents evolution of average tariff level in real terms (corrected for inflation), with 100 percent representing the level when the PPP project was initiated.

the impact of private management on various dimensions of operational efficiency found that NRW levels did not change, labor costs went up, and new expenses (such as for sludge removal) had to be incurred to comply with environmental standards. The gains came mostly from the improvement in the bill collection ratio and, to a lesser extent, from more efficient use of electricity and chemicals at wastewater treatment plants (Marin, Mas, and Palmer 2009).

## Tariffs

The tariff level is not a performance dimension in the same sense as the three dimensions reviewed previously. Whereas defining "improvements" is easy in the case of access, quality, and efficiency, making judgments about tariff levels is more difficult, because they are highly dependent on the tariff policy put in place by the government and how investments are financed. Furthermore, the fact that tariffs are rarely uniform across various categories of customers adds another element of complexity.

Although having low water tariffs may seem, at first glance, to be desirable for making piped water affordable for the poor, the actual experience in developing countries is that low tariffs have mostly benefited the connected middle class (Komives and others 2005). Keeping tariffs below cost-recovery levels could, in theory, be compensated by periodic government transfers, so that a utility can still cover its operating costs and rehabilitation or investment needs. But this rarely happens in practice, because budgetary decisions can become subverted by political agendas. As a result, many water utilities in the developing world have lacked the resources to make necessary investments for expanding the network and provide access to poor families living in periurban areas. Too often, low water tariffs have worked against the interests of the unconnected urban poor; because utilities in poor financial shape are not able to finance the investments in system expansion necessary to connect them, the poor end up having to get water from unsafe and/or more expensive sources.

The crux of the matter is that low water tariffs are not necessarily a good thing, because good service ultimately costs money. When service quality and access are low, and water tariffs are too low to cover costs and ensure the sustainability of the services, raising tariffs is often a necessary component of a utility reform irrespective of whether a private operator is being introduced. When a tariff increase happens in conjunction with a PPP, the private operator is not necessarily the cause, as is well-illustrated by the case of the management contract in Guyana (box 3.8).

Analyzing in detail the tariff changes that took place in the PPP projects under review would have gone well beyond the scope of this study. The

Box 3.8

**Raising the Water Tariff While Introducing a Private Operator in Guyana**

The case of the PPP in Guyana is a good example of why associating the introduction of a private operator with tariff increases can be misleading. The management contract started in January 2003, and a tariff increase of 37 percent was granted by the government in March 2003. At face value, the link may seem obvious, but in reality, the water utility was bankrupt. Tariff revenues did not cover operating and maintenance costs, nor even the electricity bill from the state electricity utility. The government was subsidizing the water service by not collecting the amount due to the public power utility and also had to make a budget transfer every year to ensure that employees and suppliers could be paid. The 37 percent increase was much less than was needed for the utility to cover all its operating and maintenance costs and was just a first step. Thus, the tariff rise was just one element, together with the contracting of a private operator, in a larger plan to bring the utility back to financial viability.

When the management contract started in 2003, only 45 percent of Guyana's population had access to piped water, so most of the poor who were not connected to the network did not benefit from this highly subsidized tariff. For that matter, neither did many of the customers who were connected to the network receive a benefit, because service averaged only 2.5 hours per day. Tariff increases would have been necessary under any reform proposal (whether under public or private management), if services were to become viable and sustainable.

following discussion is cast at a more general level, outlining the key factors that affect tariff levels and how they can play out in the context of a PPP project, while illustrating those points using relevant data from important projects.

### *Theory and Evidence*

The impact that the introduction of a private operator can have on tariffs depends mainly on three factors: (1) the tariff policy adopted by the government under the PPP project, which determines how much of the costs will be financed through tariff revenues, (2) the difference between tariffs and cost-recovery levels before the start of the project, and (3) the level of cost reduction the private operator can achieve through efficiency savings. In the case of concessions, the cost of private financing (and whether this cost can be offset by savings in investment efficiency) also plays a major role.

In the context of developing countries, how the last two factors combine is hard to predict. They have opposite effects and can each be very large. Initial tariff levels are often very low,[58] and whether tariffs go up or down will depend first on how close the original tariff level is to cost recovery. But on the other side, the worst-managed utilities also offer considerable scope for efficiency gains. The discussion earlier in this chapter showed that the most successful PPP projects achieved significant improvements in operational efficiency, but for cost savings to translate into lower tariffs, efficient economic regulation needs to be in place.

Given the inherent difficulties in analyzing the link between water tariffs and PPPs, it should not be surprising that the evidence available from the literature has been inconclusive. Andrés, Guasch, and others (2008) found that, in Latin America, water tariffs rose substantially in association with the implementation of PPP projects—but this is something that can be attributed largely to the fact that previous tariffs were below cost-recovery levels in many cases.

Another way to evaluate the tariff record of PPP projects is to compare tariffs between public and private utilities, but this is made difficult by the fact that public and private utilities rarely operate under the same framework. The findings of Gassner, Popov, and Pushak (2008a)—who used a very large sample of water utilities to better control for the many exogenous factors—are, therefore, of special importance. As mentioned earlier, their study compared PPP projects with corporatized public utilities—that is, utilities that tend to operate under a framework that fosters financial sustainability and full cost recovery. Thus, they tried to compare what was comparable. They found that PPPs had no statistically significant impact overall on tariff levels. The evolution and levels of water tariffs in utilities with private participation were no different statistically from those of similar public utilities. Other published evidence from country-specific data appears to be consistent with these findings, because most papers report results that are either neutral or inconclusive.

### *Evolution of Water Tariffs in Several PPPs*

Assessing the link between water PPP projects and tariff levels would require a comprehensive financial and economic analysis, which is beyond the scope of this study. This section discusses a few cases for which relevant data were

58. According to Komives and others (2005), 58 percent of utilities in the Middle East and North Africa have tariffs too low to cover basic operation and maintenance costs. Another survey of 132 utilities around the world, conducted by the European Union's Global Water Initiative in 2004, found that 39 percent operate with tariffs that do not cover such costs. The proportion of public utilities whose tariffs cover a portion of investment was even lower.

available, essentially in Western Africa and Latin America. It also discusses in more detail the experience of Manila (the Philippines), where the regulator carried out an interesting analysis of the efficiency of private operators and the effect of efficiency on tariff levels.

**Western Africa**

Many PPP projects that were put in place over the past 15 years in Western Africa inherited tariffs that were relatively high by developing-country standards, and often quite close to cost-recovery levels.[59] The evolution of average water tariffs for the PPP projects in Côte d'Ivoire, Niger (affermages), and Senegal, plus Gabon and Mali (concessions), is presented in figure 3.21. The first panel shows the evolution of the tariff in real terms (average tariff level corrected for inflation) since the entry of the private operator; the second panel allows for comparison among PPP projects on tariff levels.

In four of these five PPP projects, the average tariff level fell in real terms under private management. The fall was most notable in Gabon (to half the preconcession level by 2006) and Côte d'Ivoire (to 70 percent of the 1990 level by 2000). Customers in these two countries are now benefiting from some of the lowest water tariffs, as well as the best service levels, in Western Africa. In Senegal, the average tariff declined in real terms until the 2006 tariff adjustment, when it was raised by 15 percent, back to its pre-PPP level. But Senegal's social tariff was left untouched; in 2007, the poor households that had gained access to a connection thanks to the reform, and whose consumption was within the social lifeline of 6 cubic meters per month, were still paying an average tariff lower than the pre-PPP level.

In Niger, the average tariff went up, albeit moderately. The tariff was raised in the fourth year of the contract to increase the revenue flow for the public asset-holding company and improve the self-financing capacity of the sector; the remuneration of the private operator was not changed.[60] As in Senegal, this increase left the social tranche untouched, so the tariff paid by poor households remained below the pre-PPP level (Fall and others 2009).

59. Guinea's affermage is an exception because it started with tariffs that were well below cost-recovery levels. Tariffs were raised significantly during the 10 years of implementation, but no yearly data were available to include in figure 3.21.

60. Under Niger's affermage, tariff revenues are shared between the operator and the public asset-holding company (Société de Patrimoine des Eaux du Niger, or SPEN); only the portion of the tariff going to SPEN went up, while the private operator's volumetric fee remained unchanged.

**Figure 3.21 Evolution of Water Tariffs after Entry of Private Operators in Western Africa**

(a) Tariff index in real terms

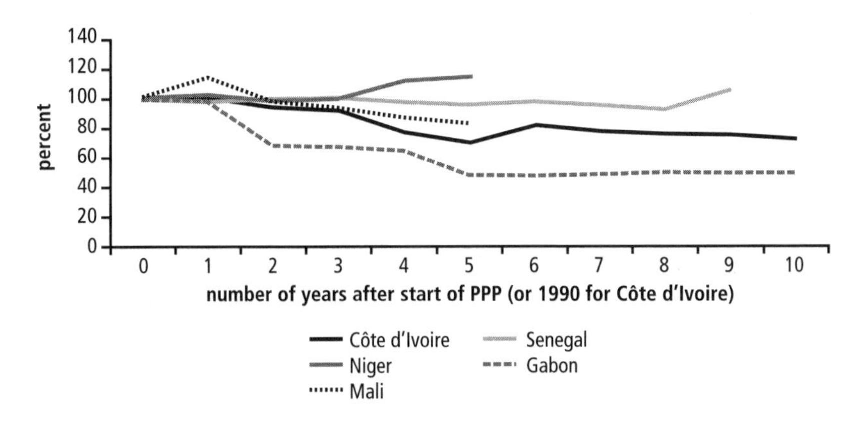


(b) Evolution of average water tariff

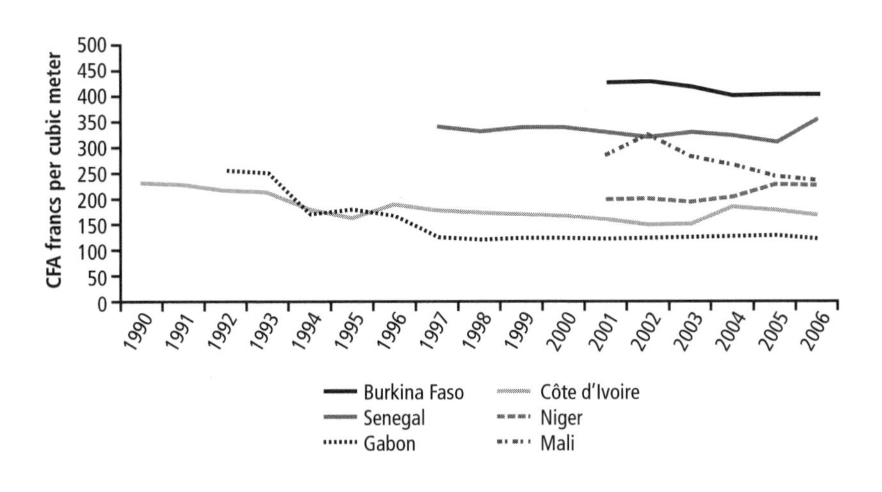


*Source:* Fall and others (2009), based on government sources.

*Note:* Water services in Burkina Faso are operated under a recently corporatized public national utility. It is included in the figure to provide some reference to a public utility operating under a framework that fosters financial viability. Water production costs in the capital, Ouagadougou, are very high, which partly explains the higher tariff levels.

**Latin America**

Latin America is the region where the greatest number of PPP projects have been implemented during the past 15 years, but assessing the impact on tariffs is difficult. In the early 1990s, several countries had just been affected by severe economic crises. Periods of hyperinflation and wide variations in exchange rates made it very hard for those countries to assess the financial situation of the utilities and calculate what tariff level would have been necessary to cover costs under a normal regime.

A large majority of the PPPs implemented in Latin America were accompanied by significant increases in water tariffs. This circumstance is not surprising, because the region had a long tradition of keeping water tariffs well below the cost-recovery level. However, the situation was exacerbated by the poor financial design of some concessions based on attempting to maximize private investment, which resulted in sharp tariff increases that often engendered major popular discontent.[61]

The evolution of the water tariff for concessionaires in Argentina, which was the largest market for private operators until 2006, is especially difficult to track. Because of the low rate of metering, customers' bills are mostly based on estimates (calculated through complex formulas), and do not reflect actual consumption volumes. Tariffs were frozen in 2001, through a government emergency order in response to the economic crisis. The most publicized case has been the concession in Greater Buenos Aires, which operated for eight years before the tariff freeze. The regulator successively awarded large tariff increases, well above the inflation level and above what had been originally planned in the contract (box 3.9). In fact, several authors have highlighted the weak regulation exercised in Argentina over water concessionaires in general and the nontransparent manner in which tariff adjustments were granted. This was probably instrumental in the current disaffection toward private concessionaires in Argentina.

In Chile, the situation has been very different. Before the regional utilities started to be transferred to the private sector in 1999, the central government had been implementing a major corporatization reform for nearly a decade. Most public water utilities were already quite efficient at the time of the transfer, and a competent national regulator had been established. Tariffs more than doubled in real terms in the decade before the transfer to private operators took place, and they continued to increase significantly after the transfer. These further increases were mostly linked to the requirement

61. This was the case, for example, in Cochabamba (Bolivia), where the tariff increase was above the amount required to reach full cost recovery. Part of the large tariff increase for residential customers was intended to finance a huge infrastructure investment that would have benefited farmers.

**Box 3.9**

**Multiple Tariff Renegotiations and Sharp Increases for Customers: The Concession in Greater Buenos Aires, Argentina**

The Buenos Aires water concession was awarded to the Aguas Argentinas consortium in 1993, on the basis of a bid that offered a reduction of 27 percent over the average tariff. In the first year of operation, the new concessionaire negotiated a first increase of 13 percent in the average tariff. Then in 1997, a major renegotiation took place, leading to an additional 19 percent tariff increase, accompanied by a major change in the tariff structure. In addition, a fixed surcharge was adopted to finance expansion in unconnected areas—to be paid by all connected customers, in a move to switch the burden of financing coverage expansion away from poor unconnected families—and to finance wastewater treatment. Ordoqui-Urcelay (2007) indicates that the water bill for households jumped by 88 percent on average between the takeover by the private operator and 2002. After the economic crisis in 2001–02, the tariff was frozen.

Average household spending on water significantly increased in Buenos Aires after the concessionaire took over. The increase has been repeatedly decried as a typical case of regulatory weakness. One major argument that has been advanced is that the introduction of the fixed surcharge was not supposed to modify the financial equilibrium of the concession, but because the concessionaire made little progress after 1998 in coverage expansion and wastewater treatment (the two priorities to be financed by the new surcharge), the surcharge may have resulted in a significant increase in the private operator's net remuneration.

to finance considerable investments in wastewater treatment and were, in general, well accepted by customers (Ducci and Medel 2007).[62]

In Colombia, Cartagena's is the only water PPP project in this study's sample whose average tariffs fell in real terms after the introduction of a private operator. In practice, the fact that water PPPs in Colombia have tended to occur mostly in the public utilities that were in the worst shape, and whose initial tariff levels were far away from cost recovery, means that tariff increases were bound to happen in most cases. Two 2007 econometric

62. Sanitation tariffs went up much more than water tariffs as a consequence of overall reform, starting at US$0.09 per cubic meter in 1990 and rising, to as much as US$0.45 per cubic meter in 2006 for the private utility in Santiago (Empresa Metropolitana de Obras Sanitarias S.A., or EMOS).

studies based on national household surveys are inconclusive regarding the impact of PPP projects on tariffs (Barrera and Olivera 2007; Gomez-Lobo and Melendez 2007).

The case of La Paz–El Alto (Bolivia) illustrates the difficulty of assessing the impact of PPP projects on tariffs. The concession was accompanied by a major overhaul of the tariff structure. Average tariffs in La Paz–El Alto increased by an average of 20 percent with the entry of the private operator in 1997, but a fixed charge that had penalized small consumers was also eliminated (Foster and Irusta 2003). In practice, this meant that most poor families with low monthly consumption saw a decline in their monthly water bills, while other customers saw sharp increases. Judging whether the 20 percent average tariff increase was justified is difficult. The average tariff of the concessionaire in La Paz–El Alto in 2003 was significantly lower than those in Cochabamba and Santa Cruz (about 21 percent and 31 percent lower, respectively), the two other large utilities in the country, but differences in local factors affecting costs (such as water availability) could explain most of these differences.

**Manila (the Philippines)**

The two concessions in Manila deserve special attention. They represent the largest population served by private operators in the developing world, and a detailed analysis of overall efficiency gains and their impact on tariffs was available from the regulator. The analysis shows that the two concessions have performed very differently in practice (Navarro 2007).

The tender process had taken place in the mid-1990s, at a time when private investors were very optimistic about the prospects for water PPPs in developing countries. The tender was based on the lowest tariff, and the results surprised even the most optimistic observers. The concessions were awarded with a considerable reduction in tariffs, with the new tariffs representing only 26.4 percent and 56.6 percent, respectively, of prebid levels in the Eastern and Western zones (Dumol 2000). The concessionaires took over in August 1997 (just one month after the start of the Asian financial crisis) and by the end of 1998, the value of the Philippine peso had halved. The financial equilibrium of the Western zone concessionaire, which carried most of the foreign currency debt of the former public utility, was severely affected.

Despite several petitions by the concessionaires, the regulator did not allow tariff adjustments above inflation to start until 2000. Then another major increase took place in 2002–03, as a result of the five-year rate-rebasing exercise that was outlined in the contract. By the end of 2006, the tariff stood at 250 percent of the pre-PPP level in the Western zone, but at only 23 percent in the Eastern zone (figure 3.22a).

Figure 3.22 **Changes in Water Tariffs over 10 Years of Concessions (Eastern and Western Zones) in Manila, the Philippines**

**(a) Evolution of water tariff in real terms**

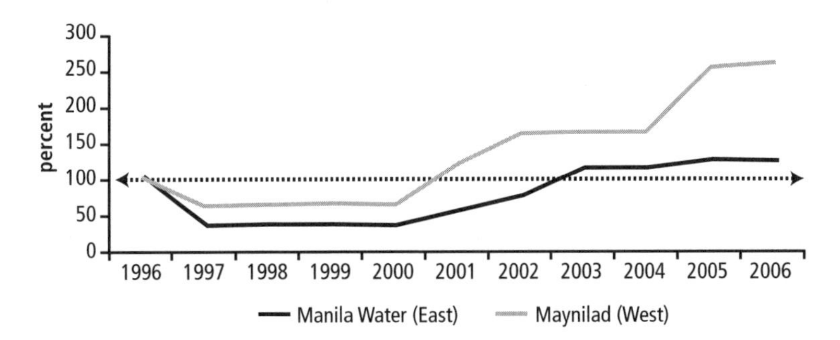


**(b) Evolution of average water tariff**

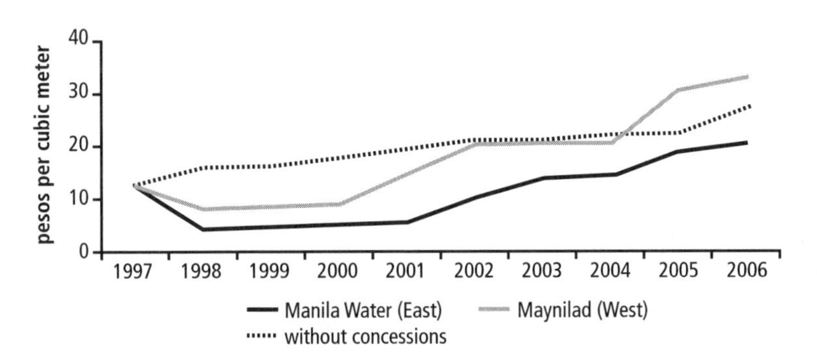


*Source:* Navarro (2007) annd Manila regulator, Metropolitan Waterworks and Sewerage Systems (MWSS).

Figure 3.22b compares the actual tariffs with the tariffs that would have been charged had Manila's water service remained in public hands. This hypothetical tariff was calculated by the regulator based on the operational efficiency of the utility under public management, taking into consideration the equivalent tariff impact of the investment actually made by the concessionaires. The results show that the tariff charged by the well-performing concessionaire in the Eastern zone was probably lower than if the services there had remained in public hands, even after the adjustments granted after 2000. It was the reverse in the Western zone.

Although these results are obviously influenced by the methodology used to establish the counterfactual (that is, how one determines the efficiency level that would have been achieved had the utility remained under public management), they do illustrate a fundamental point: ultimately, the impact of water PPPs on tariffs in the long run depends on the good or bad operational performance of the private operator. In the Eastern zone, Manila Water made sizable gains in operational efficiency, and this seemingly resulted in a lower tariff for the population, even though the tariff did go up (moderately) in real terms. In contrast, in the Western zone the concessionaire failed to achieve sizable efficiency gains. Even though a large portion of the huge tariff increase that was granted to Maynilad was linked to the higher cost of the debt that the concessionaire was carrying from the previous utility, the customers possibly ended up with a tariff that was more expensive than if the service had remained in public hands.

## Overall Performance of PPP Projects

The performance of a large number of water PPP projects in developing and transition countries was presented in the previous sections of this chapter. A consistent picture emerges despite the diversity of the projects' background and data. In many cases, private operators have improved operational efficiency, quality of service, and access to water and sanitation services. The findings can be broadly summarized as follows:

- ***Expansion of coverage.*** Although many PPP projects have expanded access to piped water, there has been a wide diversity in actual performance. The performance of PPP projects in expanding access has been highly dependent on their financial design. Expanding access to previously unserved populations often requires large investments in systems expansion, and the conditions for access to funding have varied widely among contracts. Several of the projects that relied solely on private investment achieved disappointing results.
- ***Improved service quality.*** Intermittent service is the main quality issue in water supply for most countries of the developing world. Data from both Colombia and Western Africa on the performance of several PPP projects that started under water rationing suggest that PPPs can be an efficient approach for turning around deteriorated systems.
- ***Improved operational efficiency.*** Data available on the performance of PPP projects in reducing water losses and improving bill collection and labor productivity suggest that operational efficiency is the area where the positive contribution of private operators is the most consistent.
- ***Impact on tariff levels.*** Attempting to measure the impact on tariffs of the introduction of private operators is fraught with difficulty, especially in

the context of developing countries. Published studies have not been able to find any significant difference in tariff levels between PPPs and comparable public utilities. Tariffs often went up with the implementation of PPPs, but rarely as a direct result of the entry of a private operator.

The analysis of this study has thus far treated the various dimensions of performance separately. Many projects scored well on several dimensions and deserve to be qualified overall as successes, whereas others showed some improvement for only a few indicators. The previous review of performance must now be consolidated. Identifying the partnerships that were the most successful is important not just because they show that water PPPs can work in developing countries, but also, and even more important, because they point the way for the next generation of contracts.

Obviously, assessing the overall outcome of a PPP requires some elements of judgment. To get a notion of the overall outcome of the more than 260 contracts that were signed since 1991, the study classified PPP projects in six categories (number in parenthesis) according to the following:

- PPP projects with contracts that were active by the end of 2007, with more than four years of private operation, and for which sufficient performance data were available. These projects have been classified in two categories: (category 1) contracts that were broadly successful and (category 2) contracts whose mixed outcomes (because of mixed or mediocre performance and continued difficulties in establishing a working partnership between the parties) do not allow them to be called successes.
- PPP projects for which contracts were active by the end of 2007 but for which no performance data were collected in this study and/or for which it was not possible to pass a judgment. Again, these projects have been classified in two categories: (category 3) contracts awarded since 2003 (too recent to pass a judgment) and (category 4) contracts awarded before that year (relevant for the analysis but lacking available data).
- PPP projects whose contracts were no longer in place by the end of 2007 because they were either (category 5) not renewed at expiration or (category 6) terminated early following conflicts.

The broad outcome of this classification, based on the size of the total population served by PPP projects in each category, is shown in figure 3.23 (with rounded figures). In total, about 205 million people in developing and transition countries have had experience with water PPP projects at some point since 1990. Of these, more than 160 million were still served by private operators in 2007, and about 45 million were served by utilities that had returned to public management. Each category is discussed next.

**Figure 3.23 Overall Outcomes of Water PPP Projects, by Size of Population Served, 1992–2007**

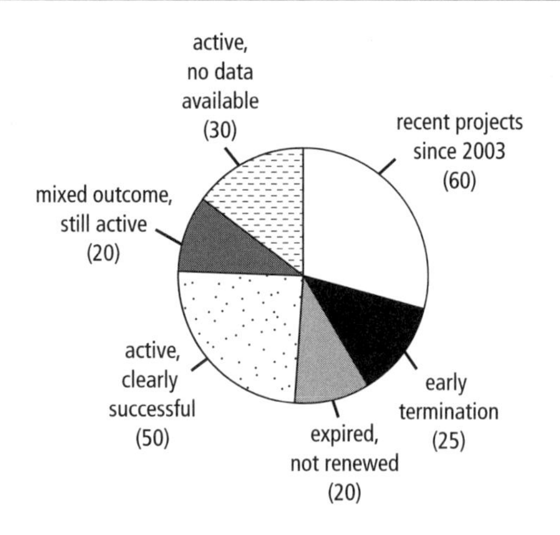


*Source:* Author's calculations.
*Note:* People in millions (rounded) are indicated in parenthesis.

The figure of 45 million for terminated or expired PPP projects means that about one-quarter of the population that was served at one time by a private operator in the past 15 years is again being served by public utilities. As many as 24 countries have returned to public management only since 1990, after having experimented with water PPP projects. These are very significant numbers, especially given the effort that international financial institutions have put into developing and financially supporting many of these arrangements. If anything, it underlines the importance of the risks and challenges associated with introducing the PPP approach in developing countries, with their volatile economic environment and often weak institutional capacities.

Passing a judgment on the performance of the PPP projects that ended with a return to public management is not straightforward, however. Several of the terminated contracts, such as those in Dar es Salaam (Tanzania), Cochabamba (Bolivia), and Buenos Aires province (Argentina), can certainly be qualified as failures. Several terminated contracts such as those in Buenos Aires (Argentina) and La Paz–El Alto (Bolivia) also brought sizable benefits for their customers, even though they proved unsustainable as the relationship between the parties deteriorated over time. In the case of management contracts, some utilities also returned to public management even though the private operator performed to the full satisfaction of the contracting government, as in Johannesburg (South Africa) (Marin, Mas, and Palmer 2009) and Gaza City (West Bank and Gaza).[63]

At the other end of the spectrum, water PPP projects that can be qualified as broadly successful serve a combined population of about 50 million (box 3.10). They include a large number of partnerships in Latin America (Argentina, Brazil, Chile, Colombia, and Guayaquil in Ecuador), Western Africa (Cote d'Ivoire, Gabon, and Senegal), Morocco, and Asia (East Manila [the Philippines] and Macao [China]). It is important to note that none of these PPP projects is perfect. All have encountered some difficulties. But overall, they have proved over time to be functioning partnerships that brought clear and consistent benefits for the population and governments concerned, validating the decision to adopt a PPP approach for reform. Most important, both public and private partners must be credited for the positive outcome in all these cases.

PPP projects that were reviewed and are still active after many years, but were classified as having a mixed outcome, represent a served population of about 20 million people (see figure 3.23). They include cases such as West Manila, Jakarta, or Maputo, where performance has been mixed or disappointing so far, often because of repeated difficulties in implementation. To be conservative, the study also included in this category several projects in which the limited data gathered suggested positive achievements but were not sufficient to pass a full judgment (Campo Grande in Brazil, Johor state in Malaysia). It is noteworthy that a successful PPP project such as that in Eastern Manila would have been classified as mixed only a few years ago. Successful reforms take time, and several PPP projects in this category might prove successful in the near future.

A large proportion of PPP projects could not be assessed, for lack of performance data. PPP contracts awarded from 2003 onward represent about 60 million people, or more than 35 percent of the population that was served by private operators in 2007. Many of these contracts are too new to pass a judgment on; they include most of those in China (serving more than 26 million people), Malaysia (7 million), Russia (10 million), and Algeria (6.5 million). This category also includes older PPP projects that are still active after more than five years of operation but for which no performance data were gathered; they include all of those in new

63. It is difficult to judge the performance of management contracts, because most of the benefits often come from intangibles such as knowledge transfer and company reorganization. These are not easily captured by performance indicators, and their impact usually goes beyond the short duration of such PPP projects.

**Box 3.10**

**A Snapshot of Successful Urban Water PPPs in Developing and Transition Countries**

In Latin America, Colombia stands out as a country where the overall experience with water PPPs has been largely positive. Gains in access, service continuity, and operational efficiency were achieved in many large and middle-size cities, often having a high poverty rate and a much-deteriorated infrastructure (Barranquilla, Cartagena, Monteria, and Soledad). In Guayaquil (Ecuador), strong progress was made in expanding access to piped water. Good results in access and service quality were also achieved by several national operators in Argentina and Brazil.

Western Africa is a region where PPP registered several notable successes (Côte d'Ivoire, Gabon, and Senegal), despite also being the region where the rate of early contract termination has been highest. Senegal is a well-known success. It has achieved the highest rate of access to piped water through household connection in the region, continuous service is provided in Dakar, the average tariff has gone down in real terms, and the urban water sector has become self-financed—a notable feat for a poor country. In Côte d'Ivoire a private operator has been operating successfully for the past 40 years, and the population with access to piped water has doubled over a decade (from 3.5 million people to about 7 million), without any direct government funding and while the average tariff went down in real terms.

*(continued)*

European Union countries (about 15 million people, including large cities such as Bucharest, Budapest, Gdansk, Prague, Sofia, and Tallinn);[64] several large contracts, such as those in Mexico (Aguascalientes, Cancun, and Saltillo), Argentina (Mendoza province), Honduras (San Pedro Sula), South Africa (Nelspruit), and Cuba (La Havana, where a foreign private operator has been operating the water utility under a mixed-ownership company

64. The performance of water PPP projects in new European Union countries was not reviewed as part of this study. Still, limited evidence available from the IBNET database (http://www.ib-net.org) suggests a positive contribution for improving operational efficiency. For example, in Prague (the Czech Republic), with 1.1 million people, during the first five years of private operation: the level of NRW went down from 17 percent to 12 percent, the bill collection ratio went up from 95 percent to 98 percent, and labor productivity improved from about 12 to 8 staff employees per 1,000 connections to about 8 staff employees.

Box 3.10

**A Snapshot of Successful Urban Water PPPs in Developing and Transition Countries *(continued)***

In Asia, the Eastern zone concession in Manila (the Philippines) has turned into a success story since the 2002 rate rebasing, and the concession in Macao (China) has been a success story for many years. In Morocco, private operators now serve close to half of the urban population, and the overall experience can be considered a success. Significant improvements were achieved in operational efficiency and service quality, and the introduction of private water operators generated strong momentum for sector reform at the national level, fostering accountability from all water providers.

Taking place in a very different context, Chile is another case in which significant benefits have been obtained with the transfer to the private sector. Contrary to what typically happened in other developing countries, the public utilities were transferred to private operators only after they had been turned into efficient providers and a competent regulatory agency had been put in place. Still, there was a huge financial benefit for the government with the transfer, which brought US$2.3 billion in cash receipts, plus a doubling of yearly revenues from income and value added taxes paid by the regional water companies (Ducci and Medel 2007). The private sector invested a total of US$1.2 billion in wastewater treatment between 1999 and 2006, and Chile is now the only country in the world (including North America and Europe) where full treatment of urban sewage is being achieved without any government money, financed entirely by private investors.

since 2000); and many contracts for small cities and towns (especially those in Colombia and Brazil).

This overall review shows that the track record of water PPPs in developing countries has been very diverse, with good, mixed, and poor projects. Such diversity in outcomes is well-illustrated by the experiences in Latin America and Sub-Saharan Africa. In these two regions, documented successful projects account for 20 million and 25 million people, respectively, compared with 16 million and 20 million people for terminated and expired projects. Clear success stories (Senegal and Cartagena [Colombia]) coexist with outright failures (Dar es Salaam [Tanzania] and Cochabamba [Bolivia]). Contract terminations have often proved disruptive for the governments and utilities involved, but the fact that 50 million people benefit from successful PPP projects also shows that this can be a viable option for developing countries, when the partnerships can be made to work. Among the various approaches available to governments for reforming their poorly performing water utilities, public-private partnership must probably be regarded more as a "high risk, high reward" option than schemes that rely on public management.

# 4.

# TOWARD MORE SUSTAINABLE WATER PPPs

This chapter summarizes a few general lessons that can be drawn from the findings and observations presented in the preceding chapters. Using this performance review of what worked or did not work in practice, some of the earlier assumptions that guided the development of water sector public-private partnerships (PPPs) in developing countries during the 1990s are questioned, especially with regard to private financing of water infrastructure. It does not pretend to offer a comprehensive discussion of all lessons to be learned from the projects that were reviewed, and most of the lessons identified would need to be researched further. Still, this discussion provides important insights on how to enhance PPP as a viable option to improve quality and access to water supply and sanitation services in the developing world.

## Lessons for More Efficient and Sustainable Water PPPs

Water PPP projects in developing countries have shown a wide diversity of outcome, and identifying the key reasons for the success or failure of specific projects can be challenging. Yet some clear patterns have emerged from this performance review, regarding (a) the appropriate main rationale for introducing private operators in support of utilities reform; and (b) the need for more realism, proper regulation, and due consideration of social issues in project design.

### *The Roles and Contributions of Private Water Operators*

The performance review suggests that the emphasis in the 1990s on implementing water PPPs in order to attract large amounts of private investment was largely misled. The largest contribution of private operators is through improved service quality and operational efficiency. Direct investment from private operators has proved disappointing in most cases. Efficient private operators do have a positive financial contribution, but it is mostly indirect, by improving the creditworthiness of the utility and allowing it, henceforth, to secure funding for investment more easily and at better terms.

**Private Operators Can Improve Service Quality and Operational Efficiency**
The performance review highlights service quality and operational efficiency as the two dimensions in which the positive contribution of private operators has been most consistent over a large number of PPP projects. In practice, though, the outcome among contracts has varied significantly. How much improvement can be achieved depends on the allocation of responsibilities and risks, which, in turn, is based on multiple factors. Two are of particular importance: (a) the incentive structure and (b) the nature of the arrangement between the private partner and the government for implementing civil works when the latter remains responsible for funding investment.

In concessions, the incentives for operational efficiency come directly from the increased profits that the private operator generates through efficiency savings. However, whether such efficiency gains can be achieved depends heavily on the capacity of the regulatory institutions that the operator faces. If regulation is lax, the operator will have little incentive to make efficiency gains and might seek, instead, to negotiate tariff increases as the easiest means to make profits.

For leases-*affermages*[65] and management contracts, incentives for efficiency are usually spelled out in more detail in the contracts. Progress is, therefore, directly related to the specific design of the remuneration and incentive scheme. In leases-affermages, the operator is typically remunerated by a fixed volumetric fee, in exchange for being responsible for all operational costs. This creates a direct incentive for making efficiency savings, because these translate directly into profits. Such a simple approach worked well in Côte d'Ivoire but proved insufficient in Guinea. The PPPs that were developed for Niger and Senegal introduced more incentives for efficiency than did typical lease-affermage contracts, with penalties for not meeting contractual targets for non-revenue water and bill collection ratios.

In management contracts, the remuneration typically comprises a fixed component plus a variable bonus for meeting contractual targets. Key elements include the choice of the indicators for measuring progress, the

65. A newly established private utility operates a publicly owned system and collects revenues that it then shares with the public owner, who remains in charge of investment.

reliability of the baseline, and the mechanism for verifying compliance with the agreed-upon targets. In Guyana, the lack of a reliable baseline and of clarity over the methodology for calculating contractual indicators created major conflicts between the parties, and was one of the factors leading to the early termination of the contract.

How the investment program is implemented is also of major importance. When private operators take over highly deteriorated systems, improvements in service quality and efficiency cannot usually be achieved without major investments in rehabilitation. The efficient and timely execution of civil works is, therefore, of paramount importance for achieving the expected performance improvements. In concessions, private operators finance and directly carry out the investment program, but in leases-affermages and management contracts, where most of the investment is not privately financed, the responsibility for the execution of the investment program is often left to the public partner. There are, however, potential risks associated with the separation of the operation and investment functions. Coordination problems may arise, and the outcome of the PPP project ends up depending, in large part, upon the ability of the public agency in charge to execute the civil works without delays. In practice, a variety of arrangements have been used for the execution of investments financed by public funds (discussed later in this chapter).

**Direct Investment by Private Operators Has Been Less Than Expected**

During the 1990s, it was widely hoped that large amounts of private investment would flow to urban water utilities, allowing donors and governments from developing countries to redirect scarce aid money to other social sectors. But these expectations did not materialize. Compared with other infrastructure sectors, the water sector experienced a very low rate of private investment.

The expectation that private investors could fund most of the investment needs of the urban water sector was largely built on the experience in Western Europe and North America, where water utilities are among the lowest credit risks. Water utilities in those regions enjoy very stable and predictable cash flows. The condition of the assets is usually well-known and so are future investment needs. Consumer access to water supply is not an issue, and tariffs are usually set at cost-recovery levels with predictable adjustment rules. For these reasons, water utilities, both public and private, have usually been able to access private debt on very favorable terms.

The situation is very different in many developing countries. Water utilities there are hardly a stable and low-risk business. Major investments are usually needed to rehabilitate deteriorated systems and expand access in a context of rapid urban development, but there is also much uncertainty as to the required level of investment. Tariffs are usually well below cost-recovery levels, and the future evolution of the regulatory framework is uncertain. Furthermore, the most solvent customers have often already invested in coping equipment (such as private wells, roof tanks, and filters) to deal with water shortages, so they resist tariff increases. Finally, there are many poor families having difficulty paying for the full cost of water.

The practical consequence is that private financiers have come to consider water concessions in most of the developing world as risky endeavors. This was apparent even during the late 1990s when international operators were attempting to raise financing for their newly awarded large concessions: private bankers have regarded nonrecourse project financing as too risky and were requiring private operators to give financial guarantees backed by their balance sheets, which, in turn, limited the amount of funds that could be invested in the sector. Furthermore, the recourse to borrowing in foreign currencies, which was used in the 1990s to compensate for the lack of local financial markets in most countries, backfired badly as the result of economic crises and turbulence in the financial markets. All in all, private equity in water has been expensive (with investors often expecting annual rates of return of 20–30 percent to compensate for risks), and so has private debt, except where creditors were able to obtain additional guarantees. This situation has put a serious constraint on the development of water concessions in many developing countries, because high rates of return for shareholders and creditors are hard to reconcile with the need to keep tariffs low and recoup investments over a long period.

### The Financial Contribution of Private Operators to Investment Has Been Mostly Indirect

Better management not only improves service, but also ultimately leads to increased investment. More efficient operation contains costs and results in more cash flow becoming available for investment. Better service makes customers more willing to pay their bills, making it easier to improve collection ratios and gradually raise tariffs to cover a larger portion of investment. All this creates a virtuous circle. Efficient operators have easier access to funding with which to expand coverage (whether from government, donors, or commercial banks), which, in turn, enlarges the customer base and augments cash flows. As Moss and others (2003) note, better management and increased investment are linked; ineffective management drives up the cost of providing services and makes it harder to obtain funding for needed investment. Any financier, whether private, a government entity, donor, customer, or taxpayer, will be reluctant to provide money if it distrusts management to deliver.

The link between better management and increased investment is well illustrated throughout the cases reviewed in this study. Only a portion of the estimated 24 million people who gained access to piped water under PPPs did so because of direct private investment, but the efficiency of private operators in turning around those utilities was, in many cases, essential for achieving this result. In Côte d'Ivoire and Gabon, for example, efficient operation allowed most investment to be funded for more than a decade through direct revenues from cash flows, without borrowing, and with average tariff levels going down in real terms. Access expansion was directly financed by customers, but this would not have been possible if the operators had not been efficient. In Senegal, the presence of a well-performing operator was instrumental in giving comfort to donors, assuring them that the investments they were financing would bring improvements for the population and that the assets would be properly maintained. The operational efficiency savings achieved in Senegal allowed the amount of revenue cash flow transferred to the public asset-holding company to go up gradually, without tariff increases, to a level where the sector gradually became self-financed. Senegal's public asset-holding company has now been able to obtain financing for investment without a sovereign guarantee.

#### *The Need for More Realistic Design and Implementation*

The review of cases indicates that water PPPs can bring tangible benefits for the populations and governments concerned. But at the same time, bringing in a private operator from outside is not a magic formula to solve all the problems of an ailing utility, following decades of mismanagement. Future PPP contracts should reflect a better understanding of what PPPs can contribute and a more adequate allocation of risks and responsibilities. Realistic expectations are essential both for setting contractual targets and for considering what should remain the responsibility of the contracting government.

### Successful Water PPPs Have Been Part of Well-Designed Overall Sector Reforms

A PPP cannot restore efficient operations and financial health in an ailing water supply utility unless it is implemented within a well-conceived, broader sectoral reform. Many elements that are essential for the viability of urban water and sanitation services are outside the control of the operator, whether public or private.

In that perspective, some important lessons can be drawn from the successful experiences of countries as diverse as Chile, Colombia, Côte d'Ivoire, Morocco, or Senegal. In all these cases, the introduction of a PPP was part of a wider reform by the central government to establish a sector framework that supported financial viability and accountability for performance. Although these countries have used a wide range of PPP schemes, all had clear policies to move to cost-recovery tariffs in a sustainable and socially acceptable manner. Also in all these cases, the financial design of the PPP ensured that sufficient funding would be available for investment in expansion and rehabilitation—an essential condition for bringing tangible benefits to the population and making the partnership sustainable. And in countries where private and public operators coexist (as in Colombia and Morocco), benchmarking was promoted at the national level to promote a sense of competition among providers. In all those cases, the government played an essential role in making the PPP projects ultimately successful, and both public and private partners deserve to be equally credited for the positive outcome.

In contrast, most of the terminated or mixed-outcome PPPs failed to solve key issues of financial viability and accountability for results. Many failed concessions were financially nonviable initially, either because of faulty design (as in Cochabamba) or because of the opportunistic behavior of some bidders. Establishing credible regulatory schemes for concessions has often been difficult, even though it has proved essential for the sustainability of PPPs. In Argentina, the country with the largest number of contract terminations, the regulatory mechanisms of most concessions proved weak and insufficient to deal with economic and political crises.

**Contracts Must Be Designed with Realistic Targets**

A recurrent problem of water PPPs has been that many contracts included unrealistic targets. For a PPP to be viable, contractual targets need to reflect realistic goals if providers are to be held accountable. In establishing such targets, governments and their advisers must not forget that many urban water utilities in the developing world operate in difficult environments. Expectations about what can be realistically achieved given each situation need to be adjusted accordingly.

Closely related to the issue of contractual targets is the difficulty of setting the baseline against which to measure progress. The original baseline used for performance measurement was often found to be inaccurate after the first months of PPP operation. This circumstance has motivated many contractual renegotiations and generated many conflicts, often seriously affecting contract implementation in the critical first years (as in Amman

[Jordan]). Establishing a reliable baseline presents a fundamental challenge: a poorly performing utility usually lacks the equipment necessary to properly measure key performance indicators, such as the level of water losses, and its operational record and customer database are usually fraught with multiple errors. Unfortunately, this problem cannot be solved just by bringing in an experienced consultant before the tender to establish the correct baseline.

A few PPP projects have solved this dilemma by adopting a flexible approach. They recognized that a reliable baseline could not be established at the time of the signing of the contract, and left to the private operator the task of putting in place the necessary framework. The baseline was agreed to in the first year of operation, under the control of an independent technical auditor. This approach was adopted in the successful management contract in Johannesburg (South Africa) (Marin, Mas, and Palmer 2009) and in the Niger affermage for establishing the non-revenue water (NRW) baseline (Fall and others 2009).[66]

**Establishing a Good Partnership That Achieves Tangible Results Takes Time**

Many PPPs that have now proved successful took a long time to deliver tangible results. For instance, the good results in Senegal took a whole decade to be achieved, and in Niger, where the same approach is being replicated, positive results are only now starting to be felt after five years of implementation. The Eastern Manila concession in the Philippines was in a tight financial situation for many years, and its performance started to improve only with the 2002 rate rebasing, more than six years after the contract award. Improving a failed urban water utility in a developing country takes time.

The outcome of a PPP depends heavily on the development of a solid collaboration between the public and private partners—something that also takes time. For the partner government, implementing a PPP often represents a radical change in administrative culture. Government officials need to move away from direct control and old habits of interfering in the operation of the water utility, to an arm's-length relationship based on contractual

66. The baseline used for non-revenue water is of paramount importance in the design of the affermages that have been put in place in Niger and Senegal, because it is part of the formula used to calculate the remuneration of the private operator. In Senegal, the baseline that was used in the original contract was a rough estimate; it proved different from the actual figure and prompted an early and difficult renegotiation. The lesson was learned in the Niger contract, which was designed so that the actual NRW baseline would be calculated in the first year of the contract, on the basis of joint monitoring by the private operator and the government, with the assistance of an independent technical auditor.

rules. This change has proved particularly difficult where the civil servants who are entrusted to act as counterparts to the private operator were the previous managers of the public utility. New regulators also need time to understand their institutional role and build capacity. The most successful PPPs were always supported by a sustained commitment from the contracting government to make the partnership work. This included being ready to adjust to changing conditions, and making the right choices when appointing key people to be involved in supervision and regulation.

A private operator brought from outside also needs time, even if it is an experienced company, to understand the situation of a specific water utility and prioritize the actions that are needed, especially if the infrastructure is severely deteriorated and few records are available. The difficulty for an outsider in mastering the hydraulics of a large and often highly deteriorated water network (usually without maps and records) can be easily underestimated. Even though there are sometimes a few opportunities for rapid improvements, achieving tangible benefits for the population usually requires important investments in rehabilitation and upgrading—a long process that involves first identifying such investments, then raising financing, and finally implementing the civil works.

**Developing Countries Can Have Their Own Private Water Operators**

In the 1990s, it was widely assumed that water PPP contracts should be signed only with private companies that had considerable previous experience in operating urban water systems. Because private operation of water utilities was by then limited to only a few countries, this conservative approach meant that most of the market was limited in practice to a few large international players.

An important finding of this review is that local private operators have developed considerably in recent years, serving more than 40 percent of the market by 2007, and several have performed well. Many of the PPPs classified as broadly successful were implemented with local private investors that had little or no previous experience in operating water utilities. In Eastern Manila (the Philippines) and Salta (Argentina), partnerships with experienced operators to transfer operational know-how allowed local operators to bridge the initial expertise gap over a few years. In other cases, investors with previous experience in the water sector through the construction, engineering, or consulting businesses proved able to operate water utilities satisfactorily (as in Brazil, Colombia, and Malaysia). Usually, those new operators just hired managers and engineers who previously worked for public utilities in order to have the necessary technical capability (as international operators do when they enter a new country).

This suggests that the need for investors to have strong previous experience in operating water utilities has probably been overestimated. The benefits from an improved commercial orientation can be achieved equally through international and local investors—and the latter have the advantage of knowing more about the local needs and culture. Operational and technical experience is important, but it can be obtained through many means. What counts is not so much whether the local investors have been involved with operating water utilities before, but whether they can credibly ensure that they have experienced people in their team to run the utility.

### *Regulation of Water PPPs*

Natural monopolies such as water utilities require economic regulation: a visible hand to substitute for the lack of market forces and ensure that the service provider does not abuse its monopolistic position. This approach is not easy to achieve in practice. Whether or not the government owns the utility, there is always a strong asymmetry of information between regulator and operator. Private operators can abuse their monopoly position to extract undue and excessive profits. Public water utilities can be captured by special interests, resulting in overstaffing, perks for political appointees, sloppy work and procurement practices, and lack of client orientation.

#### Water PPPs Are Not Necessarily More Difficult to Regulate Than Public Utilities

Concerns about the difficulty of regulating private operators, in the challenging context of developing countries, have been one of the major arguments against PPPs for urban water utilities. A regulator can, indeed, be captured by a private operator, which has strong financial interests at stake. Water PPPs are complex contracts and, in many cases, local governments with little experience in complex transactions face powerful multinationals. However, one might also observe that at least private operators operate under a framework that fosters accountability. A detailed contract spells out performance targets and mandates regular reporting. Methods for setting tariffs are stated in regulations and/or contracts, with usually much greater transparency than before. Private operators can be fined for noncompliance and can even have their contracts cancelled. Finally, PPPs tend to receive intense scrutiny from civil society—much more, in fact, than poorly performing public utilities.

#### Various Options Are Available for a Viable Regulatory System

The regulatory frameworks under which water PPPs operate in the developing world fall into two broad categories. In some countries, the focus has been on regulation by contracts, with all elements detailed in the contract and a dedicated team assigned to supervise its execution on the government's behalf. In others, the focus has been on the establishment of a broad legal and regulatory framework for the sector, usually accompanied by the creation of a regulatory agency with various degrees of discretionary power. In practice, the difference between these two approaches is not always obvious, and assessing how well they have worked would go beyond the scope of this study. Nonetheless, this review of PPP performance highlights the following observations:

Clear and detailed contracts are important, whether they are at the core or just one element of regulation. The most recent literature (Ehrhardt and others 2007), as well as this review, shows that the use of contracts as the main point of reference for holding private operators accountable has worked well in places as diverse as Western Africa, Macao (China), Colombia, the Czech Republic, and Morocco.

Anchoring the regulation of water PPPs in a comprehensive regulatory framework, in which contracts are present but the main tools are the laws and regulations, has had more mixed results, especially where newly created regulatory agencies were granted significant discretionary power. In Chile, the regulatory arrangement has worked well, but the regulator had been in place for a long time before the transfer to the private sector and was, by then, a respected and credible player. In many other places, the establishment of credible regulatory agencies has proved challenging, which eventually affected the implementation of many contracts.

The difficulties encountered with regulatory agencies are not themselves surprising: building a whole regulatory framework takes time and the process can be easily derailed. In this context, processing a single contract to address a well-identified problem may be simpler. Still, establishing a strong framework might be worth the effort, because once a framework is in place, it provides a clear and standardized point of reference that reduces the discretion that can be exercised by parties negotiating individual contracts or their adjustments. Ultimately, the right choice of whether to focus on the contract or on the framework must depend on the specifics of the country, including, among other things, the type of legal and regulatory framework (if any) that governs its water supply and sanitation sector, the current level of institutional capacity within government, and the urgency of engaging in a given partnership at the time of the decision.

### Transparency Must Be a Cornerstone of Regulation

PPPs are by nature incomplete contracts, and in the volatile environment of developing countries, it is natural that they be adjusted over time to changing conditions. However, the issue of contract renegotiation has been controversial. In many cases, it has been conducted behind closed doors, without transparency. In a comprehensive study on renegotiation of infrastructure PPPs in Latin America, Guasch (2004) found that a high proportion of water PPPs ended up being renegotiated shortly after the start of the contract.[67] All this activity fueled criticism that private operators could have been taking advantage of contractual adjustments to make financial gains, and it has undermined the credibility of the PPP approach as a valid option to improve the performance of urban water utilities in several countries.

Progress is clearly needed in this area. It is essential that the ongoing supervision and regulation of a PPP contract be carried out in a structured and fully transparent manner. All PPP contracts should be made available for public scrutiny as a standard policy. Performance monitoring and reporting of obligations must be strictly enforced and the results made available to the public as a matter of routine. Governments also have an obligation to communicate to the public the rationale and justifications behind each regulatory decision. This is especially important for all that is related to tariff adjustments (even when based on existing contractual clauses) or other modifications that may affect the financial equilibrium of a PPP. Contractual adjustments are probably unavoidable in the volatile environment of developing countries, but they cannot be expected to be accepted by the population and other stakeholders unless conducted in full transparency. This is an area where the involvement of international financial institutions during the implementation phase can be of much value, especially in countries with weak governance and institutional capacity.

### *Incorporation of Social Goals*

Social issues have been controversial in many water PPPs. It is clear that more needs to be done to ensure that more PPP projects benefit the poor. To do this, designers of PPP projects must explicitly recognize and factor in the costs of social goals as well as consider the options of subsidizing the poor and unlinking customer tariffs from the remuneration of the private operator. Also, the wide-ranging impacts of PPPs on the workforce deserve further study in order to be better addressed.

---

67. For the period 1985–2000, the study found that renegotiation in the water sector had occurred in 74 percent of cases, a much higher incidence than in other infrastructure sectors. Guasch also found that renegotiation had occurred sooner than in other sectors, taking place on average just 1.6 years after the award of the contract, and was (in most cases) instigated by the private operator. It must be noted that this study used widely defined criteria for renegotiation, and its sample included contractual arrangements that are not considered in the present study, including build, operate, and transfer (BOTs) and similar arrangements for new treatment plants.

**Water PPP Projects Need to Be Made More Pro-poor**

This study has shown that PPP projects have brought tangible benefits in access and quality of service to the population as a whole, but because data from utilities are rarely organized by customer income category,[68] the study could not specifically assess the impact of PPP projects on the urban poor.

There is circumstantial evidence that poor households significantly benefited from the increased access and reduced water rationing that was achieved by a significant number of PPP projects. This was notably the case in cities with high poverty rates, and where access was improved significantly by expanding the water network to poor neighborhoods that were previously unserved, as in Côte d'Ivoire; Senegal; Cartagena, Barranquilla, and Monteria (Colombia); Guayaquil (Ecuador); Manila (the Philippines); and even La Paz–El Alto and Buenos Aires (Argentina). In Colombia, the fact that residential customers are billed a different tariff according to their social condition allowed the study to track which household groups benefited from new connections, showing that most of the beneficiaries of access expansion were poor families.

Yet, many PPP projects do not show much evidence that sizable improvements occurred for the poor. A huge data gap exists. Admittedly, the explosive development of marginal settlements in periurban areas of many fast-growing cities has made the task of coping with demand very difficult for all utilities. Yet, given the considerable needs in the developing world, and the large efforts made by international financial institutions to support a large number of PPPs, it is fair to say that the overall outcome has been disappointing. With the exception of a few places, the benefits from water PPPs to the urban poor as a whole have not been sufficient.

Water PPP projects can bring significant benefits for a society as a whole, but the automatic trickle-down effect that was implicitly assumed in the early 1990s has not fully materialized. Private operators are merely agents acting (efficiently, it is hoped) under a set of incentives and obligations expressed in a contract, and on behalf of the contracting government. Their behavior is ultimately dictated by the design of the project. It is up to each government to ensure that efficiency gains are passed to customers and that benefits end up being equitably distributed among all groups of the society. Governments must also ensure that negative impacts are addressed through well-designed and well-funded mitigation measures.

68. For instance, data on access levels for the poor were available only in the cases of Colombia and Jakarta (Indonesia), where customers are invoiced on the basis of different tariff brackets depending on their income level. Findings from household surveys were available in only a few other cases, but their interpretation was difficult because their sample did not correspond to the boundaries of the utilities under review.

### The Cost of Social Goals Must Be Recognized in the Design of PPP Projects

One important finding from this study is that water PPP projects can provide safe water and improved sanitation services to the poorest population if the design approach differs radically from the PPP approach often adopted in the 1990s. Instead of looking at PPPs as a means to reduce the financial burden of the sector and disengage, governments need to recognize that the water sector will require continuous public support for many years to come. Reaching ambitious social goals does have a cost, and the utility's tariff revenues might not be sufficient, even allowing for potential efficiency gains brought by professional management.

The correct approach involves starting the design process by first establishing social priorities and estimating their cost. When the projected cash flows of a utility are not enough (after accounting for the expected efficiency savings) to cover the cost of such social goals, then it will be the government's task to step in with additional funding. The PPP projects that have been most successful in bringing benefits for the poor are those in which the government provided public funding in order to complement tariff revenues and accelerate progress. A strong illustration is the PPPs developed by the Colombian government under the Programa de Modernizacíon de Empresas (PME), with public grants provided to support the turnaround of utilities in poor cities with highly deteriorated infrastructure, but successful concessions in Guayaquil (Ecuador) and Cordoba (Argentina) and the affermage in Senegal also provide solid examples.

### Subsidizing of Access for the Poor Should Be Considered

In many parts of the developing world, poor families cannot be expected to afford the full cost of the water and sanitation services. Still, these are essential services to which all people must have adequate access. In addition to promoting social equity, access to piped water and adequate sanitation for the poor in dense urban areas generates large benefits for the whole population in terms of public health and environmental protection. These are reasons enough to subsidize access (and even sometimes consumption) wherever poor families cannot afford to pay the full actual costs.

Remarkably, most of the successful PPP projects included some form of subsidy scheme for the poor. A variety of mechanisms was used to make the cost of connection more affordable for poor families. Many private operators offered interest-free financing so that poor households could pay for their new connections through installments, as in Buenos Aires (Argentina), La Paz–El Alto (Bolivia), Barranquilla and Cartagena (Colombia), and Manila (the Philippines). Other PPP projects included large subsidized connection programs, as in Côte d'Ivoire, Guayaquil (Ecuador), and Senegal. The need to provide subsidized connection programs for the poor is increasingly recognized, and several projects were set up recently for existing PPP projects by providing grant money from donors through the Global Partnership for Output-Based Aid (GPOBA) facility, including in Cameroon, Morocco, and Manila (the Philippines).[69]

### Separation of Customer Tariffs from the Remuneration of the Operator Can Have Advantages

A major challenge when implementing a water reform has been how to move tariff levels toward cost recovery, starting from levels that are too low even to cover the costs of operation and maintenance (O&M). The difficulty stems from the fact that although the financial equilibrium of the operator needs to be guaranteed for the PPP project to be viable, customers are usually reluctant to accept tariff increases before tangible improvements are made—which can take many years when starting from a dilapidated infrastructure. This situation creates a major dilemma for the financial design of PPP projects.

To break the vicious circle of poor services, low willingness to pay, and insufficient revenues, designers of several PPP projects separated the remuneration to the private operator from customer tariffs. This plan allows governments or regulators to adjust tariffs more gradually, in keeping with the evolution of service quality and customers' willingness to pay. During the period until tariff levels reach full cost recovery, the government covers the operational deficit through a payment to the operator in addition to tariff revenues. The main advantage of this approach is that it allows customers to see improved services before being asked to pay for them. It also has limitations, because it can last only as long as somebody else—governments or donors—finances the revenue gap.

This approach has been used in several affermage contracts in Western Africa (Guinea, Niger, and Senegal), in which the private operator is remunerated through a volumetric fee that covers the full O&M costs but not the investment cost, which remains the responsibility of the government. The operator's fee is set by the contract and differs from the customer tariff, over which the government retains complete control. The rationale is that during the early years of the PPP project, most of the revenue collected from customers is used to cover the operator's fee, with the asset-holding company

69. GPOBA is a multidonor trust fund, administered by the World Bank, whose objective is to promote access to basic infrastructure services for the poor through the output-based aid mechanism. The grant is disbursed to the service provider only after beneficiaries' households have been connected (for more details, see http://www.gpoba.org).

having to absorb the financial gap. As the cash flow of the utility improves, thanks to efficiency gains and access expansion, a larger portion of tariff revenues can be allocated to the asset-holding company—to the point that, as in Senegal, the tariff level has now become sufficient to cover the full cost of O&M plus investment.[70]

Though delinking tariff levels from the remuneration of a private operator is relatively simple for a lease-affermage, it is more complex for a concession in which the operator is responsible not just for O&M costs but also for investment costs. Several concessions have adopted an indirect approach, with the government providing grant financing for some investments to reduce the impact on tariffs. Among the PPP projects reviewed for this study, Jakarta (Indonesia) provides the only examples of concessions in which the customer tariff was explicitly delinked from the operator's remuneration. There, the concessionaires were remunerated through a volumetric water charge (much as in an affermage) that was set in the contract, leaving the government free to change the customer tariff. Because Indonesia was severely affected by the Asian financial crisis, and because the government was unwilling to adjust tariffs for inflation during some periods, the water charge was well above the tariff level several times. The flexibility of this arrangement allowed the government to adjust its tariff policy in response to social and economic changes, and it is probably a major reason why the two Jakarta concessions have survived despite being implemented against severe odds.

### The Wide-Ranging Impact of PPPs on Labor Must Be Better Addressed

One of the major criticisms against private participation in urban water utilities has been the potentially negative impact on labor. As discussed in chapter 3, in many instances, improved labor productivity has indeed been accompanied by decreases in employment. Some massive initial layoffs took place in PPP projects in which overstaffing had been pervasive. Although they were usually justified on technical and efficiency grounds, they obviously raise serious social concerns.

Workers are assets and are, therefore, essential stakeholders in a PPP reform. Whenever layoffs are needed, adequate financial packages should be designed to mitigate the social impact on the families affected. This is necessary even though the employment changes in the water utilities may be small

70. How this approach is applied in practice varies among cases. In Senegal, the tariff level when the PPP started covered all O&M costs plus a portion of investment, but in Niger, it barely covered O&M costs. In Guinea, the initial tariff level did not even cover the full O&M cost (a World Bank loan was used to bridge the financial gap and pay the operator's fee).

in relation to total employment in the economy. In several cases of massive layoffs, such as those in Buenos Aires (Argentina), Chile, and Manila (the Philippines), workers were able to negotiate generous compensation packages, but this study could not assess whether this was the case with most other PPPs. Several PPP projects did avoid sharp drops in employment by dealing with excess staffing gradually, encouraging natural attrition and voluntary programs.

Discussions of PPPs and labor have often been limited to productivity and staffing, but the issue is much broader. Bringing in a private operator has wide-ranging implications for utility employees, who typically lose their civil servant status and its associated benefits. At the same time, the change in working conditions can introduce new opportunities, including better access to training; strengthened career paths; and, ultimately, better skilled and (it is hoped) better paid employees. Unfortunately, these important issues are still not properly understood, and further studies would be needed to assess the full impact of PPPs on labor.

## A New Generation of PPPs for Urban Water Utilities

PPPs are complex arrangements, and the developing and emerging countries are highly diverse. The study's findings suggest some key elements of an improved paradigm for water PPPs in the developing world that may help governments better harness private initiative in urban water utilities.

### *A Focus on Improving Efficiency and Service Quality*

In the 1990s, the main attraction of having private operators take over urban water utilities was their supposed ability to supply private money. Experience has shown that this was largely the wrong focus. The biggest contribution that private water operators can make is through improvements in operational efficiency and service quality. These advances, in turn, improve the creditworthiness of the utility and facilitate its access to financing for investment.

If improving operational efficiency and service quality are the key reasons for seeking PPPs, it is clear that more attention needs to be paid to providing suitable conditions for partnerships to achieve these goals. In that regard, the findings of this study largely confirm what was already known: the key ingredients are well-designed and appropriately supervised contracts, together with a sound policy and regulatory framework.[71] But the findings also highlight two important elements that have been sometimes overlooked.

71. See PPIAF and World Bank (2006) for a comprehensive review of good practice in PPP design and implementation.

First, the gains that a private operator can bring are not automatic. Private operators do not always deliver. To make progress requires careful attention to details in contract design, especially for setting up the incentive schemes. Contracts need to be well-supervised, with strict reporting requirements and, if possible, the engagement of an independent, credible technical auditor to monitor achievements. A true partnership needs to develop between the operator and the contracting government to make it easier to find solutions to the inevitable problems that will arise over time. A key challenge in this regard remains the need to build institutional capacity within the government (ministries and regulators), so that the partnership between the private operator and the public authorities can be formed on an equal footing. This is where international financial institutions can play a key role, as was shown in Amman (Jordan).[72]

Second, different types of PPPs cannot achieve the same thing. The scope for improvements rests heavily on what responsibilities are actually transferred to the private operators and on the actual duration of contracts. From this perspective, PPP projects fall into two broad groups. Management contracts are essentially low-powered instruments: the private operator takes on relatively limited risk, with limited responsibilities for a short duration. Concessions, leases-affermages, and mixed-ownership companies can be considered more as high-powered instruments, with more risks and responsibilities assumed by the private operator over a longer period.

### *Options for Securing Long-Term Financing through a Mix of Public and Private Sources*

Finding alternative ways to finance the investment needs of the urban water utilities is of major importance, given the failure overall of private operators to meet the high expectations that were placed on them during the 1990s. In practice, several financing options have been emerging, based on various combinations of public and private sources.

Though private borrowing in foreign currency to finance large water investments has often proved problematic, it would be a mistake to dismiss private financing altogether. Financial markets in some middle-income countries such as Chile, China, Colombia, Malaysia, and Morocco have matured considerably, and in the most advanced developing countries, some

72. The management contract in Amman found itself in a difficult situation after the first two years of implementation, with the public and private partners in a deadlock over several issues. At the government's request, international financial institutions supported the establishment (through grants and technical expertise) of a special project management unit with a dedicated team of professionals to ensure that the government could play a more efficient role as counterpart in the management contract.

private water companies have been able to raise medium- or even long-term funds in local currency at reasonable rates.[73] This makes water concessions a more viable proposition for the most advanced countries.

But in the majority of developing countries, whose local financial markets are not sufficiently developed, at least part of the investment carried out under water supply PPP projects will need to come from government and donor funding. This is the approach typically adopted under leases-affermages, management contracts, and most mixed-ownership companies. In such cases, the private operator's contribution to investment is indirect but can be very significant: by improving operational efficiency and service quality, the utility strengthens its ability to generate cash flow for investment, and boosts its creditworthiness, so that it can raise debt more easily (whether from donors or financial markets) and on better terms.

**Various Options Are Available for Hybrid Long-Term PPPs**

One element that came out of the study is that the traditional classification of PPP projects as management contracts, leases-affermages, and concessions has become obsolete. Most of the projects that have ultimately proved the most sustainable over time do not easily fit in one of these categories, especially when one looks at the source of financing for investment. Two good examples are the cases of Gabon and Côte d'Ivoire. In Gabon, the contractual arrangement has been a concession, whereas in Côte d'Ivoire, it has been usually classified as an affermage. Yet in both cases, most of the investment has come from directly reinvesting the cash flow collected from customer tariffs, without public or private borrowing.

In practice, the most successful financing models that are emerging for long-term water PPP projects in developing countries are basically hybrid schemes. Though the actual contractual arrangements vary, they do share the same basic principles. They transfer all commercial and operational risk to the private operator and rely for investment on a variable mix of government and private money, together with direct revenues from customer tariffs. Generally, all require some private investment (at least through equity contribution), but more as a means to ensure that the private partner has some money at stake than to cover the investment needs. Several schemes have been developed:

73. In Morocco, loan tenures of more than 16 years were available to concessionaires in 2007, which have been able to borrow about US$300 million in local currency since 1997. In Manila (the Philippines), the initial public offering (IPO) that was carried out in March 2005 for a portion of the shares of Manila Water proved to be highly successful. This was the first local listing in the Philippine Stock Exchange since the 1997 crisis to be offered simultaneously to local and international investors. It was oversubscribed 15 times and raised US$96 million.

- ***Concessions that rely largely on direct revenues from customers to finance investment.*** Concessions for combined power and water utilities in Gabon and Mali have financed a large portion of their investment programs by reinvesting each year a portion of the cash flows collected from tariff revenues, without contracting much debt. In Morocco, tariff revenues were complemented by a special surcharge allocated to a Work Fund and by a large financial contribution paid by newly connected customers (well above the actual cost of connection). The Côte d'Ivoire PPP, in which system expansion has been entirely financed by a tariff surcharge during the past 15 years without any government funding, is more of a hybrid PPP than a true affermage and also falls into this category.
- ***Affermage model in Western Africa: from public financing to self-financing through cash generation.*** This is a specific model that evolved from the French standard affermage approach, incorporating additional targets and penalties to encourage the private partner to operate efficiently. It relies heavily on public borrowing through an asset-holding company, but with the goal that after a transition period the sector should become self-financed with customer tariffs supporting both the O&M cost (represented by the operator's fee) and the government's debt service.[74] A portion of investment is left to be financed and/or implemented by the private operator.
- ***Mixed-ownership companies, with financing schemes that vary across the cases.*** This scheme is inspired by the Spanish PPP approach. Private operators initially contribute some equity funds to the utility, which then finances its investment through borrowing with or without guarantees from its main shareholders.[75] Mixed-ownership companies were successfully adopted in several large cities in Colombia (Barranquilla, Cartagena, Palmira, and Santa Marta), and in La Havana (Cuba) and Saltillo (Mexico). They are also common in several new European Union member countries (the Czech Republic and Hungary).

74. In Senegal, funding for the asset-holding company was initially obtained through government borrowing from donors at concessional rates, and the portion of the revenues left to the asset-holding company after the operator's fee had been paid was insufficient to cover the debt service. But with the gradual turnaround of the sector, the portion of tariff revenues assigned to the asset-holding company gradually increased and ended up completely covering the debt service. Thanks to its financial health, the asset-holding company has now been able to borrow directly without sovereign guarantees.

75. This approach depends on the nature of the contract between the utility and its main public shareholder, among other things (for instance, a lease in the case of Cartagena, and a concession in the case of Barranquilla), and on the creditworthiness of the utility and the conditions in local financial markets, which determine whether it can directly borrow or whether financial guarantees from the public and/or private partners are needed.

- *Concessions with government grants to finance a portion of investment.* This approach combines all three potential sources of funding (public funds, private funds, and customer tariffs). It is exemplified by the many water PPP projects developed in Colombia under the PME, with government grants to spearhead rehabilitation and expansion while reducing the impact on tariffs. It was also used in Guayaquil (Ecuador) and in Cordoba and Salta (Argentina). In Guayaquil, the connection subsidy and portion of network extension was financed by central government transfers from the telephone tax. Under the concession in Cordoba, the city government remained in charge of financing the extension of the network, and in Salta, the government provided grant financing for sewerage investment.

**Several Mechanisms Exist for Implementing Civil Works Funded with Public Money**

The arrangement between the public and private partners for the execution of the investment program plays a very important role in the final outcome of a PPP project. In cases in which most of the investment is supported by public funding, a variety of schemes have been adopted in practice. The review of cases in this study suggests that a delicate balance needs to be found between the partners. The issue is still very much open as to whether such investments are best handled by the operator or a public agency.

In Sub-Saharan Africa, the implementation of affermage contracts has typically been accompanied by the creation of a public asset-holding company, which both owns the systems and is responsible for most civil works. The track record has been mixed. In Guinea, significant delays were experienced in the implementation of the investment program, negatively affecting the performance of the PPP project. In Senegal, the asset-holding company (Sociéte Nationale des Eaux du Sénégal, or SONES) was able to execute the investment program in a timely and efficient manner—something that proved instrumental in the good overall performance of this PPP project.

Important delays in the execution of investments were also experienced with management contracts in cases in which the projects included major capital works under the responsibility of the public partner. In Amman (Jordan), the management contract had to be extended twice to allow the public agency in charge to complete the investment program. Part of the delay experienced by the public investment agency was due to cumbersome public procurement rules and coordination problems between the operator and its public counterpart.

Many successful PPP projects have been able to mitigate the risks of delays in civil works by passing on to the private operator at least some responsibilities for investment. In the affermage in Côte d'Ivoire, investment has been directly carried out by the private operator (focusing essentially on network extension). In mixed-ownership companies such as those in Cartagena, the private operator in practice runs the entire investment program, and the same occurred in management contracts in which the utilities under private management remain in charge of executing the investment program (as in Johannesburg). In Senegal and Niger, the affermage contracts include some portion of yearly rehabilitation work to be carried out directly by the operator, financed out of its own revenues. This idea was taken further in the Niger and Cameroon affermages, where the execution of some of the donor-funded investment program was delegated to the private operator to reduce its dependency on the asset-holding company's performance during the first (and most critical) years of the PPP project (box 4.1).[76]

Though passing on to the private operator the responsibility for carrying out civil works (when investment is publicly financed) may seem a solution to the problems that might result from the separation of the investment and operation, experience shows that the issue is more complex. In Guinea, the operator gradually became responsible for carrying out (directly as a construction company) a large portion of the donor-funded program, out of frustration with the underperformance of the public asset-holding company. But this undertaking distracted it from its operational responsibility; distorted the incentive framework; and, ultimately, negatively affected the performance of the PPP project. In Maputo (Mozambique), the lease contract was designed so that the private operator would be responsible for the execution of the major portion of the donor-funded investment program, but this arrangement did not prove satisfactory in practice, partly owing to the lack of experience of the operator with donor-funded investment projects.[77]

The success of the PPP project in Senegal suggests that a valid approach might be to leave large civil works to the public asset-holding company, while also passing some responsibilities to the operator, including for some civil works in network extension and rehabilitation. Delegating to a

76. In Niger, the private operator was responsible for the execution of the International Development Association (IDA)-funded network extension and subsidized connection program during the first years. In Cameroon, the affermage contract was tendered together with a construction contract for about US$20 million of civil works for emergency rehabilitation, to be implemented by the private operator in the first three years.

77. In Maputo, the contract provided for two categories of civil works to be implemented by the operator: (1) the Capital Works Projects (such as rehabilitation of mains, pumping stations, and expansion of plant production capacity) for which the operator was responsible for tendering, contracting, and supervising the work following country and donors' rules; and (2) the Delegated Works Program for the rehabilitation and extension of the distribution network, which the operator could execute itself or contract out in a more flexible manner.

Box 4.1

### How Private Operators Helped Foster Efficient Public Investment in Affermages in Senegal and Niger

The affermage schemes in Senegal and Niger illustrate the important role that an efficient private operator can play, even when the government not only finances most of the investment, but also directly implements it through an asset-holding company. Leaving these responsibilities to government is sometimes perceived as a fundamental weakness of affermages, because there is no guarantee that the government will efficiently carry out the investment function. The design of the affermage contracts in Senegal and Niger include several features to mitigate that risk.

Both contracts provide for a portion of rehabilitation investment to be carried out every year directly by the private operator. In addition to providing flexibility for the operator to carry out emergency work that directly affects operations, this arrangement allows for an ongoing benchmarking of procurement costs. Because the private operator has some responsibility for investment, many items such as pipes are purchased by both the operator and the public agency, providing a valuable reference point. A recent study in Niger showed that the private operator was able to purchase pipes and meters at prices 30 percent to 75 percent lower than those paid by the public utility before the affermage.

The private operator in Senegal also plays a major role in defining the investment program, even though it is not responsible for most of the program's execution. The contract provides for revision of the government's investment program every three years, based on the operator's input. This approach ensures that the money is spent on operational priorities that have a real impact on the service delivered to the population, instead of on white elephants of dubious operational value. Besides shaping investment decisions, the operator is also routinely consulted in the supervision of all civil works. In addition, it has strong incentives to ensure that these are carried out adequately, because it will have to operate the assets afterwards. Most important, the operator must give its own approval before any civil work is approved and paid for by the government, giving the staff of the asset-holding company a strong incentive to properly supervise the work of the contractors.

private operator the execution of some donor-funded emergency works—as a way to mitigate the risk of nonperformance by a newly created public asset-holding company—might be an option to consider, but with caution. Given this study's findings, how best to channel public funds for investment remains an open question. The best solution probably depends on each case.

## Time to Rebalance the Debate

In retrospect, it is not surprising that PPPs for urban water utilities have been controversial. In the 1990s the challenges in the sector were huge. Major structural reforms were needed in many countries, and it was inevitable that such reforms would prove difficult and encounter stiff resistance from those who benefited from the status quo. By considering public-private partnership as the preferred (and sometimes only) option for reforms in many countries, stakeholders generated unduly high expectations that translated into contractual obligations that were difficult to meet. Disappointment was bound to result.

This study has shown that soundly designed and implemented PPP projects can be a viable option for improving the operational and financial performance of poorly performing water utilities in developing countries. But contrary to initial expectations, long-term PPP projects in the water sector did not bring in the large international inflows of capital that were needed to close the access gap and rehabilitate ailing utilities. The main contribution of private operators has been in improving service quality and operational efficiency, something they achieved to varying degrees over a large number of projects. Admittedly, this report is not the last word on the many issues related to water PPP, and much remains to be done in terms of analysis. This is especially the case with regard to the impact on the poor, which could not be analyzed in detail.

As these lessons were being internalized by the market, a new generation of water PPP projects has been gradually emerging, based on a more balanced allocation of risks between the public and private partners. This new generation of PPP projects focuses on using private operators to improve service quality and operational efficiency, rather than to attract private money. Although governments retain the main responsibility for mobilizing long-term financing, several financing options are emerging, including hybrids with some contributions from the private sector. The recent surge of new private operators from developing countries is also a major development that may greatly enhance the viability and acceptability of water PPPs in many countries.

This study showed that public-private partnership can be a viable option to improve urban water utilities in developing countries, but this does not mean that it should be the only option on the table. Well-managed public utilities can be found within the developing world—a fact that was obscured during the PPP fashion of the 1990s. Several public reforms supported by the World Bank and other donors in the past decade have enjoyed success, even in very poor places such as Burkina Faso; Uganda; and Cambodia's capital, Phnom Penh. Many well-managed public water utilities are operating in many countries, and sector practitioners are gaining a better understanding of how to successfully reform water utilities (Baietti, Kingdom, and van Ginneken 2006). Well-functioning public utilities are those in which tariffs are high enough to recover costs, labor productivity is satisfactory, customers pay their bills, and the infrastructure is properly maintained and efficiently run. To achieve this, the concerned governments have had to make choices. These include putting in place sound tariff policies, refraining from interference in operations, and putting in place professional management that is held accountable for results.

Among the various options for water utilities reform, it is probably fair to characterize PPPs as high-risk, high-reward propositions for governments, the more so as one moves across the PPP spectrum from management contracts to leases-affermages and then to concessions. PPP projects are complex arrangements; they are difficult to implement in the context of developing countries' weak institutional capacity and economic volatility and involve significant transaction costs. They are vulnerable to vested interests and, unfortunately, can make easy targets for opportunistic politicians, especially during the early years when the population often still does not perceive tangible improvements in service. And finally, the fact that private operators do not always deliver must not be overlooked.

Public-sector-only reforms, by contrast, might bring less-rapid improvements but might also pose much less political risk. However, as the experiences of successful public utilities have shown, these reforms are no less demanding for governments. To achieve tangible results, the reforms require the same structural changes in the sector in general, and in the governance of the utility in particular, as do PPP projects. Vested interests that are opposed to reform must still be dealt with. Applying sound commercial management practices is needed to ensure the financial sustainability of the services.

Although this study has shown that private participation in the urban water sector brought significant benefits to tens of millions of people in many countries, the benefits brought about by PPPs probably go well beyond the populations served by these successful projects. Even in a country where PPP projects serve only a minority of water customers, the introduction of a few private operators alongside public utilities can bring a sense of competition to an erstwhile monopolistic sector, increasing the pressure on public players to improve performance and deliver better services. This situation is well-exemplified by the recent positive evolution of the urban water sector in countries such as Brazil, Colombia, and Morocco. Also, the well-publicized difficulties encountered by private operators in meeting contractual targets for expansion of coverage in periurban areas probably helped bring to the top of donors' agendas the key issue of how to provide access to basic water and sanitation service to the urban poor.

It is time to move beyond a rather sterile public-versus-private debate. That an international private operator has been running the water utility of La Havana in Cuba since 2000 shows that the question of how to turn around poorly performing urban water utilities cannot be reduced to mere ideology. Well-run public utilities of the developing world have much in common with efficient private providers. The type of water and sanitation services that a country wants and how much it is prepared to pay for them (through tariffs and tax transfers) are fundamental decisions that any government needs to face. Part of the allure of PPPs in the 1990s was that they appeared to offer an elegant solution to a complex problem. However, the fact that the underlying causes of the sector's woes were social and political as much as technical and financial was not sufficiently appreciated at that time. Setting tariff levels, holding service providers accountable, and making financial trade-offs to achieve competing social and environmental goals are public issues that are independent of how utilities are managed, and they cannot be avoided by public authorities. Private operators can inject dynamism, improve efficiency, and build a culture of service and accountability, but they cannot obviate the need for the government to make profound and sometimes difficult choices. At root, many countries are still struggling with the principle that was adopted at the Dublin Water Summit in 1992—that water is both a social and an economic good. The difficulty of finding an institutional model that delivers that good in a socially and economically viable way reveals deep underlying issues that are not yet resolved among stakeholders. The question of how to solve this challenge in each country goes well beyond the mere question of whether water utilities should be privately or publicly managed.

Though wiser contractual arrangements are emerging for PPP projects in water and sanitation—with better risk allocation; sounder financial design; and a larger, more diverse corps of operators to rely upon—delegating the provision of urban water supply and sanitation services to private operators remains a challenging endeavor, and seems unlikely at this time to become the dominant model in the sector. This said, the scope for private sector involvement has been widening. Many water utilities that have remained public have opened the door to the private sector without delegating their management responsibilities. The public utility in Bogotá (Colombia) subcontracted the entire operation of its network to local private companies. In Mexico City, private operators have been handling meter reading and billing for more than a decade. A growing number of public utilities are expanding subcontracting in a drive to gain flexibility and contain costs, following a practice well-developed by efficient public utilities in Northern Europe.

Private financing is also making new headway, although not through private operators themselves but through private financiers, which are showing a growing interest in directly financing well-performing public utilities, without guarantees from central governments. This approach is becoming possible in more and more countries as local financial markets have developed and public utilities have improved their performance and creditworthiness. In addition to accessing private funds through build, operate, and transfer (BOT) contracts for treatment facilities, an increasing number of public utilities have started to access private financing, either by selling shares to minority shareholders (as in São Paulo, Brazil) or by borrowing directly from local private banks (as in Colombia, Mexico, and Morocco). In many countries, increased access to market-based financing without sovereign guarantees provides incentives for public water utilities to improve their financial and operating performance—in turn, helping them to compete on more equal terms with privately managed utilities.

The private sector has much to offer, and in many forms. There is no such thing as a purely public water utility, because most of them routinely involve the private sector in multiple ways (such as with civil works and service contractors or with commercial banks). Similarly, purely private water utilities (such as in Chile) represent only a very small portion of PPPs, and arrangements such as those the PPP in Senegal involve government agencies as much as the private sector. Recently, publicly owned utilities from both developed and developing countries have started to look for contracts outside their jurisdictions, where they act legally as private entities. As the traditional boundaries between public and private sectors are getting blurred, decision makers in government have an increasing number of options at their disposal to improve the performance of their urban water utilities. To tackle the immense challenges facing the urban water sector in developing countries, policy makers need all the help they can get. It might just be time for a broader partnership, one that includes all and excludes none.

## Appendix A

# Water PPPs Whose Performance Was Reviewed under This Study

**Table A.1 Water PPPs Whose Performance Was Reviewed under This Study**

| Economy | Type of PPP | Performance dimension | Years of operation | Population served (millions)[a] | Data source |
|---|---|---|---|---|---|
| **Africa** | | | | | |
| Chad | MC | S, OE | 2000–04 | 1.1 | Bank (SCS) |
| Côte d'Ivoire | Lease-affermage | A, S, OE, T | 1961– | 8.7 | Gov, Op (SCS) |
| Gabon | Concession, MC | A, S, OE, T | 1997– | 0.7 | Gov, Op (SCS) |
| Guinea | Lease-affermage | A, S, OE | 1991–98 | 1.1 | Bank (SCS) |
| Mali | Concession | A, OE, T | 2000–05 | 1.5 | Gov (SCS) |
| Mozambique | | | | | |
| Beira and three others | MC | S, OE | 1999– | 0.6 | Bank |
| Maputo | Lease-affermage | A, S, OE | 1999– | 1.0 | Bank |
| Niger | Lease-affermage | A, S, OE, T | 2001– | 1.8 | Gov, Op (SCS) |
| Senegal | Lease-affermage | A, S, OE, T | 1996– | 4.7 | Gov, Op (SCS) |
| South Africa | | | | | |
| Dolphin Coast | Concession | OE (limited data) | 1999– | 0.05 | Palmer Development Group (2003) |
| Johannesburg | MC | OE | 2001–06 | 3.2 | Gov, Op (SCS) |
| Queenstown | Lease-affermage | OE | 1992– | 0.2 | Palmer Development Group (2003), Op |
| Stutterheim | Lease-affermage | OE (limited data) | 1995–01 | 0.05 | Plummer (2000) |
| Uganda | | | | | |
| Kampala | MC | S, OE | 2002–04 | 0.6 | Gov (SCS) |
| Zambia | | | | | |
| Zambia mining towns | MC | S, OE | 2001–04 | 0.3 | Bank |

| | | | | | |
|---|---|---|---|---|---|
| **East Asia and Pacific** | | | | | |
| China | | | | | |
| Macao | Concession | OE, T | 198x– | 0.6 | Op |
| Indonesia | | | | | |
| Batam Island | Concession | OE (limited data) | 1995– | 0.7 | Op |
| East Jakarta | Concession | A, OE | 1998– | 3.1 | Gov, Op |
| West Jakarta | Concession | A, OE | 1998– | 3.1 | Gov, Op |
| Malaysia | | | | | |
| Johor province | Concession | OE | 2001– | 3.0 | Op |
| Philippines | | | | | |
| East Manila | Concession | A, S, OE, T | 1996– | 5.4 | Gov, Op (SCS) |
| West Manila | Concession | A, S, OE, T | 1996– | 6.4 | Gov, Op (SCS) |
| **Europe and Central Asia** | | | | | |
| Albania | | | | | |
| Durres and others | MC | S, OE | 2003– | 0.8 | Bank |
| Armenia | | | | | |
| Yerevan | MC | S, OE | 2000– | 1.3 | Bank (SCS) |
| Kosovo | | | | | |
| Gjakova and Rahovec | MC | S, OE | 2002–05 | 0.2 | Bank (SCS) |
| Turkey | | | | | |
| Antalya | Lease-affermage | S (limited data) | 1996–02 | 0.6 | Bank |

*(continued)*

**Table A.1 Water PPPs Whose Performance Was Reviewed under This Study *(continued)***

| Economy | Type of PPP | Performance dimension | Years of operation | Population served (millions)[a] | Data source |
|---|---|---|---|---|---|
| **Latin America and the Caribbean** | | | | | |
| Argentina | | | | | |
| Buenos Aires | Concession | A, S, OE, T | 1993–06 | 8.0 | Pap, Op |
| Cordoba | Concession | A, S, OE | 1997– | 1.3 | Pap, Op |
| Corrientes province | Concession | A, S, OE, T | 1991– | 0.6 | Op (SCS) |
| La Rioja province | Concession, MC | A, S, OE, T | 1999– | 0.2 | Op (SCS) |
| Salta province | Concession | A, S, OE, T | 1998– | 1.2 | Op, Gov (SCS) |
| Santa Fe province | Concession | A, S, OE | 1995–06 | 1.9 | Pap, Op |
| Bolivia | | | | | |
| La Paz–El Alto | Concession | A, S, OE, T | 1997–07 | 1.5 | Pap, Gov |
| Brazil | | | | | |
| Campo Grande | Concession | OE | 2000– | 0.8 | IBNET |
| Campos | Concession | A, OE | 1999– | 0.4 | IBNET |
| Itapemirim | Concession | A, OE | 1998– | 0.2 | IBNET |
| Limeira | Concession | A, OE | 1995– | 0.3 | IBNET, Op |
| Manaus | Concession | A, OE | 2000– | 1.5 | IBNET, Op |
| Niteroi | Concession | A, OE | 1999– | 0.5 | IBNET |
| Paranagua | Concession | A, OE | 1997– | 0.15 | IBNET |
| Petropolis | Concession | A, OE | 1998– | 0.25 | IBNET |
| Prolagos (Cabo Frio) | Concession | OE | 1998– | 0.3 | IBNET |
| Tocantins state | Divestiture | OE | 1999– | 1.0 | IBNET |

|  |  |  |  |  |  |
|---|---|---|---|---|---|
| Chile |  |  |  |  |  |
| ESSBIO-ESSEL | Divestiture | S, OE, T | 1999– | 2.1 | Gov (SCS) |
| Los Lagos (ESSAL) | Divestiture | S, OE, T | 1999– | 0.6 | Gov (SCS) |
| Santiago (EMOS) | Divestiture | S, OE, T | 1999– | 5.5 | Gov (SCS) |
| Valparaiso (ESVAL) | Divestiture | S, OE, T | 1999– | 1.4 | Gov (SCS) |
| Colombia |  |  |  |  |  |
| Barranquilla | Mixed-ownership | A, S, OE, T | 1997– | 1.3 | Gov, Op (SCS) |
| Cartagena | Mixed-ownership | A, S, OE, T | 1996– | 0.9 | Gov, Op (SCS) |
| Girardot | Mixed-ownership | A, S, OE, T | 1999– | 0.15 | Gov |
| Marinilla and neighboring towns | Concession | A, S, OE | 1997– | 0.25 | Gov, Op |
| Monteria | Concession | A, S, OE, T | 2000– | 0.3 | Gov, Op, Pap |
| Palmira | Mixed-ownership | A, S, OE, T | 1998– | 0.25 | Gov |
| Santa Marta | Mixed-ownership | A, S, OE, T | 1997– | 0.4 | Gov, Op |
| Soledad | Concession | A, S, OE, T | 2001– | 0.45 | Gov, Op |
| Tunja | Concession | A, S, OE, T | 1996– | 0.15 | Gov |
| Ecuador |  |  |  |  |  |
| Guayaquil | Concession | A, S, OE, T | 2001– | 2.2 | Op, Gov (SCS) |
| Guyana | MC | S, OE | 2002–06 | 0.4 | Bank |
| Trinidad | MC | S, OE | 1996–99 | 1.2 | Bank |
| Venezuela, R. B. de |  |  |  |  |  |
| Lara | MC | S, OE | 1999–02 | 1.3 | Gov (SCS) |
| Monagas | MC | S, OE | 1997–01 | 0.6 | Gov (SCS) |

*(continued)*

**Table A.1 Water PPPs Whose Performance Was Reviewed under This Study *(continued)***

| Economy | Type of PPP | Performance dimension | Years of operation | Population served (millions)[a] | Data source |
|---|---|---|---|---|---|
| **Middle East and North Africa** | | | | | |
| Jordan | | | | | |
| Amman | MC | S, OE | 1999–06 | 2.2 | Gov (SCS) |
| Lebanon | | | | | |
| Tripoli | MC | OE (limited data) | 2003–05 | 0.2 | Pap |
| Morocco | | | | | |
| Casablanca | Concession | A, OE | 1997– | 3.8 | Gov, Op |
| Rabat | Concession | OE | 1999– | 2.0 | Gov |
| Tangiers and Tetouan | Concession | A, OE | 2000– | 1.1 | Gov |
| West Bank and Gaza | | | | | |
| Gaza City | MC | S, OE | 1996–03 | 1.0 | Bank |
| **Total** | | | | 98.4 million | |

*Source:* Author.

*Note:* The table shows contracts that have been in place for at least five years (three for management contracts) and for which performance data were obtained as part of this study. It does not include large PPPs, such as those in Cochabamba (Bolivia), Dar es Salaam (Tanzania), and Tucuman (Argentina), that were terminated after just one or two years of private operation (these PPPs are still discussed in the text of the report). The estimates of population served take into account the corresponding coverage levels. No end-year means that the PPP project was still active by the end of 2007.

A = access; Bank = World Bank project documents; Gov = government; IBNET = International Benchmarking Network for Water and Sanitation Utilities, http://www.ib-net.org; MC = management contract; OE = operational efficiency; Op = operator; Pap = published paper(s); S = service quality; SCS = specific case study prepared with consultant(s) as part of this review; T = Tariff.

a. 2007 or last year of PPP.

## Appendix B

# New Connections and Increased Access in 36 Large PPP Projects

**Table B.1 New Connections and Increased Access in 36 Large PPP Projects**

| PPP project | Period of reference | New water connections | Population gaining access to piped water |
|---|---|---|---|
| Manila, Eastern zone (Philippines) | 1997–2006 | 250,000 | 2,900,000 |
| Manila, Western zone (Philippines) | 1997–2006 | 230,000 | 1,900,000 |
| Jakarta, West and East (Indonesia) | 1998–2006 | 210,000 | 2,000,000 |
| Batam Island (Indonesia) | 1996–2006 | 80,000 | 500,000 |
| Johor state (Malaysia) | 2000–06 | 180,000 | 800,000 |
| Macao (China) | 1991–2006 | 75,000 | 180,000 |
| Casablanca (Morocco) | 1997–2005 | 260,000 | 1,200,000 |
| Rabat (Morocco) | 2002–05 | 65,000 | 250,000 |
| Tangiers and Tetouan (Morocco) | 2002–05 | 45,000 | 150,000 |
| Gabon | 1996–2006 | 50,000 | 300,000 |
| Mali | 2001–05 | 40,000 | 400,000 |
| Buenos Aires (Argentina) | 1993–99 | 240,000 | 2,000,000 |
| Corrientes, La Rioja, and Salta provinces (Argentina) | 1991 or 1998–2006 | 140,000 | 650,000 |
| Guayaquil (Ecuador) | 2001–06 | 160,000 | 800,000 |
| Santa Fe (Argentina) | 1995–2006 | 60,000 | 500,000 |
| Cordoba (Argentina) | 1997–2006 | n.a. | 200,000 |
| La Paz–El Alto (Bolivia) | 1997–2005 | 80,000 | 400,000 |
| Tocantins state (Brazil) | 1999–2006 | 130,000 | 600,000 |
| Manaus (Brazil) | 2000–06 | 50,000 | 300,000 |
| Campos, Niteroi, and Petropolis (Brazil) | 1999–2006 | 80,000 | 350,000 |

| | | | |
|---|---|---|---|
| Barranquilla, Santa Marta, and Soledad (Colombia) | 1997–2006 | 100,000 | 600,000 |
| Monteria and Tunja (Colombia) | 1996 or 2000–05 | n.a. | 200,000 |
| **Total for 30 large concessions** | | **2,500,000** | **17,200,000** |
| Cartagena (Colombia) | 1996–2006 | 70,000 | 500,000 |
| Guinea | 1989–98 | n.a. | 600,000 |
| Senegal | 1996–2006 | 190,000 | 1,700,000 |
| Côte d'Ivoire | 1990–2006 | 300,000 | 4,000,000 |
| Maputo (Mozambique) | 1999–2006 | 20,000 | 150,000 |
| Niger | 2001–07 | 30,000 | 450,000 |
| **Total for 6 large leases-affermages** | | **600,000** | **7,500,000** |
| **TOTAL** | | **3,200,000** | **24,700,000** |

*Source:* Author's calculation (rounded figures) based on various sources.

*Note:* n.a. = not applicable.

# BIBLIOGRAPHY


Abdala, Manuel. 1997. *Welfare Effects of Buenos Aires Water and Sewerage Services Privatization.* Washington, DC: World Bank.

———. 2001. "Institutions, Contracts, and Regulation of Infrastructure in Argentina." *Journal of Applied Economics* 4 (2).

Agrawal, Pronita. 2006. "Improving WSS Services through Private Sector Partnerships: Jamshedpur Utilities." Water and Sanitation Program (WSP) field note, June, World Bank, Washington, DC.

Aït Ouyahia, Meriem. 2006. "Public-Private Partnerships for Funding Municipal Drinking Water Infrastructure: What Are the Challenges?" Discussion Paper, Policy Research Institute of Canada, Ottawa.

Alcazar, Lorena, Manuel Abdala, and Mary Shirley. 2000. "The Buenos Aires Water Concession." Policy Research Working Paper 2311, World Bank, Washington, DC.

Andrés, Luis, Maktar Diop, and José Luis Guasch. 2008. "Achievements and Challenges of Private Participation in Infrastructure in Latin America: Evaluation and Future Prospects." In *Euromoney Infrastructure Financing,* ed. Henry Davis. Oxford, U.K.: Oxford University Press.

Andrés, Luis, José Luis Guasch, Thomas Haven, and Vivien Foster. 2008. *The Impact of Private Sector Participation in Infrastructure: Lights, Shadows and the Road Ahead.* Latin American Development Forum Series. Washington, DC: World Bank.

Anwandter, Lars, and Teofilo Ozuna. 2002. "Can Public Sector Reforms Improve the Efficiency of Public Water Utilities?" *Environmental and Development Economics* 7: 687–700.

Armstrong E., M. Robinson, and S. Hoy. 1976. *History of Public Waterworks in the United States 1776–1976.* Chicago: American Public Works Association.

Artana, Daniel, Fernando Navajas, and Santiago Urbiztondo. 1999. "Governance and Regulation: A Tale of Two Concessions in Argentina." In *Spilled Water: Institutional Commitment in the Provision of Water Services,* ed. Pablo Spiller, 197–248. Washington, DC: Inter-American Development Bank.

Aspiazu, Daniel. 2005. *Privatización del sistema de agua potable y saneamiento en el área metropolitana de Buenos Aires: Debilidad institucional, regulatoria y enseñanzas.* Santiago, Chile: GWP-CEPAL.

Aspiazu, Daniel, M. Schorr, E. Crenzel, G. Forte, and J. C. Marín. 2005. "Agua potable y saneamiento en Argentina, privatizaciones, crisis, inequidades e incertidumbres futura," *Cuadernos del CENDES* 22 (59): 45–67.

Auriol, Emmanuelle, and Aymeric Blanc. 2007. "Public Private Partnerships for Water and Electricity in Africa." Working Paper 38, Agence Française de Développement, Paris.

Avendaño, Ruben. 2006. "Concesiones posibles en municipios: El caso de Montería en Colombia." *AquaVitae Magazine* 21 (2): 25–28.

Avendaño, Ruben, and Federico Basañes. 1999. "Private Participation at the Subnational Level: Water and Sewerage Services in Colombia." In *Can Privatization Deliver?* ed. Federico Basañes and others, 217–44. Washington, DC: Inter-American Development Bank.

Baietti, A., W. Kingdom, and M. van Ginneken. 2006. "Characteristics of Well-Performing Public Water Utilities." Water Supply and Sanitation Working Note 9, World Bank, Washington, DC.

Baietti, Aldo, and Peter Raymond. 2005. *Financing Water Supply and Sanitation Investments: Utilizing Risk-Mitigation Instruments to Bridge the Financing Gap.* Washington, DC: World Bank.

Ballance, Tony, and Andrew Taylor. 2005. *Competition and Economic Regulation in Water: The Future of the European Water Industry.* London: IWA Publishing.

Ballance, Tony, and Sophie Trémolet. 2005. *Private Sector Participation in Urban Water Supply in Sub-Saharan Africa.* Frankfurt, Germany: KfW and GTZ (German Development Cooperation).

Barja, G., D. McKenzie, and M. Urquiola. 2005. "Bolivian Capitalization and Privatization: An Approximation to an Evaluation." In *Reality Check: The Distributional Impact of Privatization in Developing Countries,* ed. J. Nellis and N. Birdsall. Washington, DC: Center for Global Development.

Barja, G., and M. Urquiola. 2003. "Capitalization and Privatization in Bolivia: An Approximation to an Evaluation." Center for Global Development, Washington, DC.

Barrera, Felipe, and Mauricio Olivera. 2007. "Does Society Win or Lose as a Result of Privatization? Provision of Public Services and Welfare of the Poor: The Case of Water Sector Privatization in Colombia." Research Network Working Paper R-525, Latin American Research Network, Inter-American Development Bank, Washington, DC.

Basañes, Federico, Evamaria Uribe, and Robert Willig. 1999. *Can Privatization Deliver?* Washington, DC: Inter-American Development Bank.

Basañes, Federico, and Robert Willig. 2002. *Second-Generation Reforms in Infrastructure Services.* Washington, DC: Inter-American Development Bank.

Bertolini, Lorenzo. 2006. "How to Improve Regulatory Transparency: Emerging Lessons from an International Assessment." Gridlines 11, PPIAF, Washington, DC.

Birdsall, Nancy, and John Nellis. 2002. "Winners and Losers: Assessing the Distributional Impact of Privatization." Working Paper 6, Center for Global Development, Washington, DC.

Bitran, Gabriel, and Pamela Arellano. 2005. "Regulating Water Services: Sending the Right Signals to Utilities in Chile." Public Policy for the Private Sector Series Note 286, World Bank, Washington, DC.

Bitran, Gabriel, and E. Valenzuela. 2003. "Water Services in Chile: Comparing Public and Private Performance." Public Policy for the Private Sector Series Note 255, World Bank, Washington, DC.

Blanc, A., and C. Guesquières. 2006. "Secteur de l'eau au Sénégal: Un partenariat équilibré entre acteurs publics et privés pour server les plus démunis." Working Paper 24, Agence Française de Développement, Paris.

———. 2006. "Decentralization and the Free Basic Water Policy in South Africa: What Role for the Private Sector?" Working Paper 25, AFD (French Development Agency), Paris.

Boccanfuso, Dorothée, Antonio Estache, and Luc Savard. 2006. *Water Reforms in Senegal, a Regional and Interpersonal Distributional Impact Analysis.* Québec, Canada: Sherbrooke University.

Boda, Zsolt, Gabor Scheiring, Emanuele Lobina, and David Hall. 2008. "The Case of Hungary." In *Social Policies and Private Sector Participation in Water Supply: Beyond Regulation*, ed. Naren Prasad, 178–202. New York: Palgrave Macmillan.

Botton, Sarah. 2007. *Mettre en Place une Démarche d'Ingénierie Sociale: Suez et les Quartiers Défavorisés de Buenos Aires.* Paris: Ecole Nationale des Ponts et Chaussées.

Brailowsky, Alexandre. 2007. *L'impossibilité de travailler dans une logique d'intérêts partagés: Une approche transversale des contrats de La Paz et de Buenos Aires.* Paris: Suez.

Brenneman, Adam, and Michel Kerf. 2002. "Infrastructure and Poverty Linkages: A Literature Review." Mimeo, December, World Bank, Washington, DC.

Breuil, Lise. 2004. "Renouveler le partenariat public-privé pour les services d'eau dans les pays en développement." PhD diss., Ecole Nationale du Génie Rural et des Eaux et Forêts (ENGREF), Paris.

Brocklehurst, Clarissa, and Jan Janssens. 2004. "Innovative Contracts, Sound Relationships: Urban Water Sector Reform in Senegal." Water and Sanitation Sector Board Discussion Paper 1, World Bank, Washington, DC.

Brook, Penelope. 1999. "Lessons from the Guinea Water Lease." Public Policy for the Private Sector Series Note 78, World Bank, Washington, DC.

Brook, Penelope, and Alain Locussol. 2001. *Easing Tariff Increases: Financing the Transition to Cost-Covering Tariffs in Guinea.* Washington, DC: World Bank.

Brook, Penelope, and Suzanne Smith, eds. 2001. *Contracting for Public Services: Output-Based Aid and Its Applications.* Washington, DC: World Bank.

Budds, Jessica, and Gordon McGranaham. 2003a. "Are the Debates on Water Privatization Missing the Point? Experiences from Africa, Asia, and Latin America." *Environment and Development*, 15 (2).

———. 2003b. "Privatization and the Provision of Urban Water and Sanitation in Africa, Asia, and Latin America." Working Paper, International Institute for Economic Development, London.

Burgos, Mejia, and Francisco Felix. 2005. *Seguimiento a la participación del sector privado en acueducto y alcantarillado realizada desde el programa de modernización empresarial.* Bogota: Department of National Planning, Government of Colombia.

Camdessus Panel (World Panel on Financing Water Infrastructure). 2005. *Financing Water for All: Report of the Camdessus Panel.* World Water Council, 3rd Water Forum. http://www.financingwaterforall.org.

Casarín, Ariel, Jose Delfino, and Maria Eugenia Delfino. 2007. "Failures in Water Reform: Lessons from the Buenos Aires Concession." *Utilities Policy* 15 (4): 234–47.

Castalia. 2006. "Case Studies on Water and Sanitation Sector Economic Regulation: Themes from Four Case Studies (Vanuatu, Manila, Senegal, and Colombia)." Final Report to the World Bank, Consultant report, Washington, DC.

Chisari, O., and A. Estache. 1999. "Universal Service Obligations in Utility Concession Contracts and the Needs of the Poor in Argentina's Privatization." Policy Research Working Paper 2250, World Bank, Washington, DC.

Chisari, O., A. Estache, and C. Romero. 1999. "Winners and Losers from Utility Privatization in Argentina, Lessons from a General Equilibrium Model." *World Bank Economic Review* 13 (2): 357–78.

Chisari, Omar, Antonio Estache, and Catherine Waddams Price. 2001. "Access by the Poor in Latin America's Utility Reforms: Subsidies and Service Obligations." Discussion Paper 2001–75, UN-WIDER (United Nations University, World Institute for Development Economic Research), Helsinki, Finland.

Clarke Annez, Patricia. 2006. "Urban Infrastructure Finance from Private Operators: What Have We Learned from Recent Past Experiences?" Policy Research Working Paper 4045, World Bank, Washington, DC.

Clarke, George, K. Kosec, and S. Wallsten. 2004. "Has Private Participation in Water and Sewerage Improved Coverage? Empirical Evidence from Latin America." Policy Research Working Paper 3445, World Bank, Washington, DC.

Clarke, George, Claude Ménard, and Ana Maria Zuluaga. 2002. "Measuring the Welfare Effects of Reform: Urban Water Supply in Guinea." *World Development* 30 (9): 1517–37.

Clarke, G., and S. Wallsten. 2002. "Universal(ly bad) Service: Providing Infrastructure Services to Rural and Poor Urban Consumers." Policy Research Working Paper 2868, World Bank, Washington, DC.

Cohen, Michael, Alexandre Brailowsky, and Barbara Chénot-Camus. 2004. *Citizenship and Governability: The Unexpected Challenges of the Water and Sanitation Concession in Buenos Aires*. New York: New School University.

Collignon, Bernard. 2002. *Urban Water Supply Innovations in Côte d'Ivoire: How Cross-Subsidies Help the Poor.* Nairobi, Kenya: Africa Water and Sanitation Program.

Constance, Paul. 2006. "When a Water Meter Is Worth More Than a House: The Case of Barranquilla in Colombia." *IDB America Magazine*. Washington, DC: Inter-American Development Bank.

Cours des Comptes. 2003. *La gestion des services publics de l'eau et de l'assainissement en France*. Paris. http://www.ccomptes.fr/CC/documents/RPT/RapportEau.pdf.

Crampes, Claude, and Antonio Estache. 1996. "Regulating Water Concessions: Lessons from the Buenos Aires Concession." Policy Research Working Paper 2311, World Bank, Washington, DC.

Cranfield University and DFID (U.K. Department for International Development). 2004. *Regulating Public Private Partnerships for the Poor, Urban Water Supply, Buenos Aires Case Study*. London: DFID.

Crocker, Keith, and Scott Masten. 2000. "Prospects for Private Water Provision in Developing Countries: Lessons from 19th-Century America." Mimeo. University of Michigan Business School, Ann Arbor, and World Bank.

Dardenne, Bertrand. 2006. "The Role of the Private Sector in the Peri-Urban or Rural Water Services in Emerging Countries." Global Forum on Sustainable Development, OECD, Paris, November.

Davis, Jennifer. 2004. "Corruption in Public Service Delivery: Experience from South Asia's Water and Sanitation Sector." *World Development* 32 (1): 53–71.

Delfino, Jose, Ariel Casarín, and Maria Eugenia Delfino. 2007. "How Far Does It Go? The Buenos Aires Water Concession a Decade After the Reform." Social Policy and Development Program Paper 32, United Nations Research Institute for Social Development, Geneva.

Diaz, Javier. 2003. *La participatión del sector privado en los servicios de agua potable y saneamiento en San Pedro Sula*. Honduras: Inter-American Development Bank.

Drees-Gross, Franz, Jordan Schwartz, Maria-Angelica Sotomayor, and Alexander Bakalian. 2005. *Output-Based Aid in Water: Lessons in Implementation from a Pilot in Paraguay*. Washington, DC: World Bank.

Ducci, Jorge. 2007. *Salida de operadores internacionales de agua en América Latina*. Washington, DC: Inter-American Development Bank.

Ducci, Jorge, and Omar Medel. 2007. "Chilean Experience with Private Sector Participation in Water Utilities." Mimeo, September. ETWWA, World Bank, Washington, DC.

Dumol, Mark. 2000. *The Manila Water Concession: A Key Government Official's Diary of the World's Largest Water Privatization*. World Bank Directions in Development Series. Washington, DC: World Bank.

Eberhard, Anton. 2006a. *The Independence and Accountability of Africa's Infrastructure Regulators: Reassessing Regulatory Design and Performance.* South Africa: University of Cape Town.

———. 2006b. "Matching Regulatory Design to Country Circumstances: The Potential of Hybrid and Transitional Models." Gridlines 23, PPIAF, Washington, DC.

———. 2007. "Infrastructure Regulation in Developing Countries: An Exploration of Hybrid and Transitional Models." Working Paper 4, PPIAF, Washington, DC.

Ehrhardt, David, Eric Groom, Jonathan Halpern, and Seini O'Connor. 2007. "Economic Regulation of Urban Water and Sanitation Services: Some Practical Lessons." Water Sector Board Discussion Paper Series 9, World Bank, Washington, DC.

El-Nasser, Hazim. 2007. "Case Study of the Management Contract for Amman Water and Wastewater Services." Mimeo, World Bank: Washington, DC.

Esguerra, Jude. 2005. "Manila Water Privatization: Universal Service Coverage after the Crisis?" Working paper, June, UNRISD, Geneva.

Estache, Antonio. 2005a. "PPI Partnerships vs. PPI Divorces in Developing Countries." Policy Research Working Paper 3470, World Bank, Washington, DC.

———. 2005b. "Privatization in Latin America: The Good, the Ugly and the Unfair." Mimeo, World Bank, Washington, DC.

Estache, Antonio, Vivien Foster, and Quentin Wodon. 2002. *Accounting for Poverty in Infrastructure Reform: Learning from Latin America's Experience.* Washington, DC: World Bank.

Estache, Antonio, José Luis Guasch, and L. Trujillo. 2003. "Price Caps, Efficiency Pay-offs, and Infrastructure Contract Renegotiation in Latin America." Policy Research Working Paper 3129, World Bank, Washington, DC.

Estache, Antonio, and E. Kouassi. 2002. "Sector Organization, Governance, and the Inefficiency of African Water Utilities." Policy Research Working Paper 2890, World Bank, Washington, DC.

Estache, Antonio, and M. Rossi. 1999. "Comparing the Performance of Public and Private Water Companies in Asia and Pacific Region: What a Stochastic Costs Frontier Shows." Policy Research Working Paper 2152, World Bank Institute, Washington, DC.

———. 2002. "How Different Is the Efficiency of State and Private Water Companies in Asia?" *World Bank Economic Review* 16 (1): 139–48.

Estache, Antonio, and Lourdes Trujillo. 2003. "Efficiency Effects of Privatization in Argentina Water and Sanitation Services." *Water Policy* 5: 369–80.

Fall, Matar, Philippe Marin, Alain Locussol, and Richard Verspyck. 2009. "Reforming Urban Water Utilities in Western and Central Africa: Experiences with Public Private Partnerships." Water Sector Board Discussion Paper Series 13, World Bank, Washington, DC.

Foster, Vivien. 2005. "Ten Years of Water Services Reforms in Latin America: Towards an Anglo-French Model." Water and Sanitation Sector Board Discussion Paper 3, World Bank, Washington, DC.

Foster, V., and O. Irusta. 2003. "Does Infrastructure Reform Work for the Poor? A Case Study on the Cities of La Paz and El Alto in Bolivia." Policy Research Working Paper 3177, World Bank, Washington, DC.

Foster, Vivien, A. Gomez Lobo, and Jonathan Halpern. 2000. *Designing Direct Subsidies for Water and Sanitation Services, Panama Case Study.* LCSPR, Universidad de Chile, Santiago.

Franceys, Richard. 2008. "GATS, 'Privatisation' and Institutional Development for Urban Water Provision: Future Postponed?" *Progress in Development Studies* 8 (1): 45–58.

Franceys, Richard, and A. Nickson. 2003. *Tapping the Market—The Challenge of Institutional Reform in the Urban Water Sector.* Palgrave Macmillan.

Franceys, Richard, and A. Weitz. 2003. "Public-Private Community Partnerships in Infrastructure for the Poor." *Journal of International Development* 15: 1083–98.

Galiani, S., P. Gertler, and E. Schargrodsky. 2005. "Water for Life: The Impact of the Privatization of Water Services on Child Mortality." *Journal of Political Economy* 113: 83–120.

Gassner, Katharina, Alexander Popov, and Nataliya Pushak. 2008a. *Does Private Sector Participation Improve Performance in Electricity and Water Distribution? An Empirical Assessment in Developing and Transition Countries.* PPIAF Trends and Policies Series. Washington, DC: PPIAF.

———. 2008b. "Does the Private Sector Deliver on its Promises? Evidence from a Global Study in Water and Electricity." Gridlines 36, PPIAF, Washington, DC.

Gentry, Bradford, and Auyuan Alethea. 2000. *Global Trend in Urban Water Supply and Wastewater Financing and Management: Changing Roles for the Public and Private Sector.* Paris: OECD.

Gleick, Peter, Gary Wolff, Elizabeth Chalecki, and Rachel Reyes. 2002. *The New Economy of Water: The Risks and Benefits of Globalization and Privatization of Fresh Water.* Oakland, CA: Pacific Institute.

Gökgür, Nilgün, and Leroy Jones. 2006. "Privatization of Mozambique Water Sector." Boston Institute for Developing Economies (BIDE), Boston, MA.

Gomez-Lobo, Andrés. 2001. *Making Water Affordable: Output-Based Consumption Subsidies in Chile.* Washington, DC: World Bank.

Gomez-Lobo, A., and D. Contreras. 2003. "Water Subsidy Policies: A Comparison of the Chilean and Colombian Schemes." *World Bank Economic Review* 17 (3): 391–407.

Gomez-Lobo, Andrés, Vivien Foster, and Jonathan Halpern. 2000. "Information and Modeling Issues in Designing Water and Sanitation Subsidy Schemes." Policy Research Working Paper 2345, World Bank, Washington, DC.

Gomez-Lobo, Andrés, and M. Melendez. 2007. "Social Policy, Regulation and Private Sector Participation: the Case of Colombia." Working Paper, UNRISD, Geneva.

Goueti Bi, A. T. 1996. "Private Company and State Partnership in the Management of Water Supply and Sewage Services: The Case of SODECI in Côte d'Ivoire." In *Managing Water Resources for Large Cities and Towns, Report of the Beijing Water Conference,* 90–102. Nairobi, Kenya: United Nations Center for Human Settlements (UN-Habitat).

Groom, Eric, Jonathan Halpern, and David Ehrhardt. 2006. "Explanatory Notes on Key Topics in the Regulation of Water and Sanitation Services." Water Sector Board Discussion Paper Series 6, World Bank, Washington, DC.

Guasch, José Luis. 2002. *Concessions of Infrastructure Services: Incidence and Determination of Renegotiations—An Empirical Evaluation and Guidelines for Optimal Concession Design.* Washington, DC: World Bank.

———. 2004. "Granting and Renegotiating Infrastructure Concession: Doing It Right." WBI Development Series, World Bank Institute, Washington, DC.

Guasch, José Luis, Jean-Jacques Laffont, and Stéphane Straub. 2003. "Renegotiation of Concession Contracts in Latin America." Policy Research Working Paper 3011, World Bank, Washington, DC.

Guislain, Pierre, and Michel Kerf. 1996. "Concessions: The Way to Privatize Infrastructure Sector Monopolies." Public Policy for the Private Sector Series Note 59, World Bank, Washington, DC.

Gulyani, Sumila, D. Talukdar, and R. M. Kariuki. 2005. "Water for the Urban Poor: Water Markets, Household Demand, and Service Preferences in Kenya." Water Supply and Sanitation Sector Board Discussion Paper Series 5, World Bank, Washington, DC.

Gupta, Pankaj, Ranjit Lamech, Farida Mazhar, and Joseph Wright. 2002. "Mitigating Regulatory Risk for Distribution Privatization—The Wrld Bank Partial Risk Guarantee." Energy and Mining Sector Board Discussion Paper, World Bank, Washington, DC.

Gurria Task Force (Task Force on Financing Water for All). 2006. Gurria Panel Report. http://www.financingwaterforall.org/index.php?id=1097.

Haarmeyer, David, and Ashoka Mody. 1998. "Financing Water and Sanitation Projects—The Unique Risks." Public Policy for the Private Sector Note 151, World Bank, Washington, DC.

Haggarty, Luke, Penelope Brook, and Ana Maria Zuluaga. 2002. "Water Services Contracts in Mexico City." In *Thirsting for Efficiency: The Economics and Politics of Urban Water System Reform,* ed. Mary Shirley, 139–88. Oxford, U.K.: Pergamon Press.

Hall, David, and Emanuele Lobina. 2006. *Pipe Dreams: The Failure of the Private Sector to Invest in Water Services in Developing Countries.* London: Public Services International Research Unit (PSIRU), Greenwich University and Word Development Movement.

Hall, David, Emanuele Lobina, and Robin De La Motte. 2005. "Public Resistance to Privatization in Water and Energy." *Development Practice* 15 (3) and 15 (4).

Harris, Clive. 2003. "Private Participation in Infrastructure in Developing Countries: Trends, Impacts, and Policy Lessons." Working paper, World Bank, Washington, DC.

Harris, Clive, John Hodges, Michael Schur, and Padmesh Shukla. 2003. "Infrastructure Projects: A Review of Cancelled Projects." Note 252, Private Sector and Infrastructure Network, World Bank, Washington, DC.

Hemson, David, and Herbert Batidzirai. 2002. *Public Private Partnerships and the Poor: Dolphin Coast Water Concession (South Africa)*, WEDC (Water Engineering Development Center), Loughborough University, Leicestershire, U.K.

Hibou, B., and O. Vallée. 2007. "Energie du Mali ou les paradoxes d'un échec retentissant." Working Paper 37, Agence Française de Développement, Paris.

Hodge, Graeme. 2000. *Privatization: An International Review of Performance, Theoretical Lenses on Public Policy.* Boulder, CO: Westview Press.

Hume Smith, William. 2006. "Improving the Regulation of Water and Sanitation Services: Preliminary Review to Categorize, Describe and Assess Incentive Provisions in Management Contracts." Mimeo, World Bank, Washington, DC.

IBNET (International Benchmarking Network for Water and Sanitation Utilities). http://www.ib-net.org.

IDB (Inter-American Development Bank). 2003. *Financing Water and Sanitation Services in Latin America and the Caribbean.* Washington, DC: IDB.

Idelovitch, Emanuel, and Klas Ringskog. 1995. *Private Sector Participation in Water Supply and Sanitation in Latin America, Directions in Development.* Washington, DC: World Bank.

Jacobson, Charles D., and Joel A. Tarr. 1995. "Ownership and Financing of Infrastructure: Historical Perspectives." Policy Research Working Paper 1466, World Bank, Washington, DC.

Jammal, Yahya, and Leroy Jones. 2006. *Impact of Privatization in Africa: Uganda Water.* Boston, MA: Boston Institute for Development Economics (BIDE).

Jensen, Olivia, and Frederic Blanc-Brude. 2006a. "The Handshake: Why Do Governments and Firms Sign Private Sector Participation Deals? Evidence from the Water and Sanitation Sector in Developing Countries." Policy Research Working Paper, World Bank, Washington, DC.

———. 2006b. "The Institutional Determinants of Private Sector Participation in the Water and Sanitation Sector in Developing Countries." Policy Research Working Paper, World Bank, Washington, DC.

Jme'an, Suhail, and Khairy Al-Jamal. 2004. "Non-Revenue Water in Gaza." Presentation for World Bank Training Workshop on Non-Revenue Water Reduction, World Bank, Washington, DC, June.

Kariuki, Mukami, and Jordan Schwartz. 2005. "Small-Scale Service Providers of Water Supply and Electricity." Policy Research Working Paper 3727, World Bank, Washington, DC.

Kauffmann, Céline, and Edouard Pérard. 2007. *Stocktaking of the Water and Sanitation Sector and Private Sector Involvement in Selected African Countries.* Background Note for NEPAD regional roundtable, OECD, Paris, France.

Kerf, Michel. 1998. "Concessions for Infrastructure: A Guide to Their Design and Award." Technical Paper 399, World Bank, Washington, DC.

———. 2000. "Do State Holding Companies Facilitate Private Participation in the Water Sector? Evidence from Côte d'Ivoire, The Gambia, Guinea, and Senegal." Policy Research Working Paper, World Bank, Washington, DC.

Kerf, Michel, and Ada Karina Izaguirre. 2007. "Review of Private Participation in Developing Country Infrastructure." Gridlines 16, PPIAF, Washington, DC.

Kingdom, William, Roland Liemberger, and Philippe Marin. 2006. "The Challenge of Reducing Non-Revenue Water in Developing Countries, How the Private Sector Can Help: A Look at Performance-Based Service Contracting." Water Supply and Sanitation Sector Board Discussion Paper Series 8, World Bank, Washington, DC.

Kirkpatrick, Colin, and David Parker. 2005. *Domestic Regulation and the WTO: The Case of Water Services in Developing Countries*. Oxford, U.K.: Blackwell Publishing.

Kirkpatrick, C., C. Parker, and Y. Zhang. 2004. "State versus Private Sector Provision of Water Services in Africa: A Statistical, DEA and Stochastic Cost Frontier Analysis." Working Paper 70, Working Paper Series, Centre on Regulation and Competition, Institute for Development Policy and Management, University of Manchester, U.K.

Klein, Michael. 1996. "Economic Regulation of Water Companies." Policy Research Working Paper 1649, World Bank, Washington, DC.

Klein, Michael, and Neil Roger. 1994. "Back to the Future: The Potential in Infrastructure Privatization." Public Policy for the Private Sector Series Note 30, World Bank, Washington, DC.

Komives, Kristin. 1999. "Designing Pro-Poor Water and Sewer Concessions: Early Lessons from Bolivia." Policy Research Working Paper, World Bank, Washington, DC.

Komives, Kristin, and Penelope Brook. 1998. "Expanding Water and Sanitation Services to Low-income Households: The Case of the La Paz–El Alto Concession." Viewpoint Note 178, World Bank, Washington, DC.

Komives, Kristin, Vivien Foster, Jonathan Halpern, and Quentin Wodon. 2005. *Water, Electricity, and the Poor: Who Benefits from Utility Subsidies?* Washington, DC: World Bank.

Krause, Matthias. 2009. *The Political Economy of Water and Sanitation*. New York: Routledge.

Lauria, Donald, Omar Hopkins, and Sylvie Debomy. 2005. "Pro-Poor Subsidies for Water Connections in West Africa." Water Supply and Sanitation Working Note 3, World Bank, Washington, DC.

Lee, Cassey. 2008. "The Case of Malaysia." In *Social Policies and Private Sector Participation in Water Supply: Beyond Regulation*, ed. Naren Prasad, 149–77. New York: Palgrave Macmillan.

Lobina, Emanuele, and David Hall. 2003. *Problems with Private Water Concessions: A Review of Experience*. London: Public Services International Research Unit (PSIRU), Greenwich University.

Lorrain, Dominique. 1992. "The French Model of Urban Services." *West European Politics* 15 (2): 77–92.

Maceira, Daniel, Pedro Kremer, and Hilary Finucane. 2007. *El desigual acceso a los servicios de agua corriente y cloacas en la Argentina*. Buenos Aires: Centro de Implementación de Políticas Públicas para la Equidad y el Crecimiento.

Mandri-Perrott, Cledan. 2009. "Public and Private Partnerships in the Water and Wastewater Sector—Developing Sustainable Legal Mechanisms." Water Law and Policy Series, IWA Publishing, London.

Manzetti, Luigi, and Carlos Rufin. 2005. "Private Utility Supply in a Hostile Environment: The Experience of Water/Sanitation and Electricity Distribution Utilities in Northern Colombia, Dominican Republic, and Ecuador." Working Paper, Inter-American Development Bank, Washington, DC.

Marin, Philippe, Jean Pierre Mas, and Ian Palmer. 2009. "Using a Private Operator to Establish a Corporatized Public Utility: The Management Contract for Johannesburg Water." Water Sector Board Working Notes 20, World Bank, Washington, DC.

Marin, Philippe, and Ada Karina Izaguirre. 2006. "Private Participation in Water: Towards a New Generation of Projects?" Gridlines 14, PPIAF, Washington, DC.

Marin, Philippe, Jean Pierre Mas, and Ian Palmer. 2009. "The Johannesburg Management Contract." Working Note 20, World Bank, Washington, DC.

Mariño, Manuel, Jack Stein, and Francisco Wulff. 1998. "Management Contracts and Water Utilities: The Case of Monagas State in Venezuela." Public Policy for the Private Sector Series Note 166, World Bank, Washington, DC.

Martinand, Claude. 2007. *Water and Sanitation Services in Cities and Countries Bordering the Mediterranean Sea*. Paris: Institut de la Gestion Déléguée.

Maslyukivska, Olena, M. Sohail, and B. Gentry. 2003. "Private Sector Participation in the Water Sector in the ECA Region: Emerging Lessons." Report prepared for the World Bank and OECD, Washington, DC.

McGranahan, Gordon, and David Satterthwaite. 2006. *Governance and Getting the Private Sector to Provide Better Water and Sanitation Services to the Urban Poor.* London: International Institute for Environment and Development.

McKenzie, D., and D. Mookherjee. 2003. "The Distributive Impact of Privatization in Latin America: Evidence from Four Countries." *Economia* 3 (2): 161–233.

Megginson, W., and J. Netter. 2001. "From State to Market: A Survey of Empirical Studies on Privatization." *Journal of Economic Literature* 39: 321–89.

Ménard, Claude, and George Clarke. 2000. "A Transitory Regime: Water Supply in Conakry, Guinea." In *Thirsting for Efficiency: The Economics and Politics of Urban Water System Reform,* ed. Mary Shirley, 273–316. Oxford, U.K.: Pergamon Press.

———. 2002. "Reforming Water Supply in Abidjan, Côte d'Ivoire: A Mild Reform in a Turbulent Environment." In *Thirsting for Efficiency: The Economics and Politics of Urban Water System Reform,* ed. Mary Shirley, 233–72. Oxford, U.K.: Pergamon Press.

Ménard, Claude, and S. Saussier. 2000. "Contractual Choice and Performance: The Case of Water Supply in France." In *The Economics of Water Contracts: Theory and Applications,* ed. Eric Brousseau and Jean-Michel Glachan, 440–62. Cambridge, U.K.: Cambridge University Press.

Mohajeri, S., B. Knothe, D-N. Lamothe, and J-A. Faby. 2003. *Aqualibrium: European Water Management between Regulation and Competition*. Brussels: European Commission.

Moss, Jack, Gary Wolff, Graham Gladden, and Eric Guttierrez. 2003. "Valuing Water for Better Governance: How to Promote Dialogue to Balance Social, Environmental and Economic Values." Business and Industry CEO Panel, March.

Mugabi, Josses, and Philippe Marin. 2008. "PPPs in Urban Water: Lessons from Yerevan, Armenia." *Management, Procurement and Law* 161 (November).

Mugisha, Silver, William Muhairwe, Josses Mugabi, and Philippe Marin. 2007. "Transforming Public Water Utilities through Private Sector-Like Management Principles: The Case of Uganda." Mimeo, World Bank, Washington, DC.

Nankani, Helen. 1997. "Testing the Waters: A Phased Approach to a Water Concession in Trinidad and Tobago." Public Policy for the Private Sector Series Note 103, World Bank, Washington, DC.

National Research Council. 2002. *Privatization of Water Services in the United States: An Assessment of Issues and Experience*. Washington, DC: National Academy Press.

Navarro, Mariles. 2007. "A Case Study of the Experience of the Two Water Concessions in Manila, Its Impact on the Population and Improved Operational Efficiency (1997–2006)." Mimeo, July, PPIAF and World Bank, Washington, DC.

Nellis, John. 2007. *Privatizing Basic Utilities in Africa: A Rejoinder*. Brasilia, Brazil: UNDP International Poverty Center.

Nickson, Andrew. 2001a. "The Córdoba Water Concession in Argentina." Working Paper 442–05, April, GHK International, London.

———. 2001b. "Establishing and Implementing a Joint Venture: Water and Sanitation Services in Cartagena, Colombia." Working Paper 442–03, January. GHK International, London.

Nickson, Andrew, and Claudia Vargas. 2002. "The Limitations of Water Regulation: The Failure of the Cochabamba Concession in Bolivia," *Bulletin of Latin American Research* 21 (1): 99–120.

Nouha, Hassan, Mehdi Berradi, Mohamed Dinia, and Mustapha El Habti. 2002. "Partenariats publics privés: Cas de la distribution d'eau potable au Maroc." Paper presented at international workshop in Amman, Jordan, October.

OECD (Organisation for Economic Co-operation and Development). 2008. "Public Private Partnerships: in Pursuit of Risk Sharing and Value for Money." OECD, Paris.

Ordoqui-Urcelay, Maria Begoña. 2007. *Servicios de agua potable y alcantarillado en la ciudad de Buenos Aires: Factores determinantes de la sustentabilidad y el desempeño*. CEPAL Series 126. Santiago de Chile: CEPAL.

Palaniappan, Meena, Heather Cooley, and Peter Gleick. 2006. "Assessing the Long-Term Outlook for Current Business Models in the Construction and Provision of Water Infrastructure and Services." Global Forum on Sustainable Development, OECD, Paris, November.

Palmer Development Group. 2003. *Water PPPs in South Africa and Their Impact on the Poor.* Report prepared for the Department of Water Affairs and Forestry, Government of South Africa.

Pan-American Health Organization (PAHO). http://www.paho.org.

Parker, David, and Colin Kirkpatrick. 2003. "Privatization in Developing Countries: A Review of the Evidence and the Policy Lessons." Working Paper Series 55, Center on Regulation and Competition, Manchester, U.K.

Pérard, Edouard. 2008. "Private Sector Participation and Regulatory Reform in Water Supply: The Southern Mediterranean Experience." Working Paper 265, OECD Development Center, Paris.

Pezon, Christelle. 2000. *Le service d'eau potable en France de 1850 à 1995.* Paris: Presses du Conservatoire National des Arts et Métiers.

Piaget, J. 2003. "Limits in Water Concession Contracts: The Case of Aguas de Aconquija (Argentina)." Master's thesis, University of Lausanne (HEC), Switzerland.

Plane, Patrick. 1999. "Privatization, Technical Efficiency, and Welfare Consequences: The Case of the Côte d'Ivoire Electricity Company." *World Development* 27 (2): 343–60.

Plantz, Daniel, and Frank Schroeder. 2007. "Moving Beyond the Privatization Debate: Different Approaches to Financing Water and Electricity in Developing Countries." Occasional Paper 34, Friedrich Ebert Stiftung, Berlin.

Plummer, Janelle. 2000. *Favourable Policy and Forgotten Contract: Private Sector Participation in Water and Sanitation Services in Stutterheim, South Africa.* GHK International and University of Birmingham, U.K.

PPI Projects Database (Private Participation in Infrastructure Projects Database). World Bank/PPIAF. http://ppi.worldbank.org/resources/ppi_aboutDb.aspx.

PPIAF (Public-Private Infrastructure Advisory Facility). 2004. *Labor Issues in Infrastructure Reforms: A Toolkit.* http://www.ppiaf.org/LaborToolkit/toolkit.html.

PPIAF and World Bank. 2001. *Toolkit: A Guide for Hiring and Managing Advisors for Private Participation in Infrastructure.* Washington, DC: PPIAF and World Bank.

———. 2006. *Approaches to Private Participation in Water Services, A Toolkit.* Washington, DC: World Bank.

Prasad, Naren, ed. 2008. *Social Policies and Private Sector Participation in Water Supply: Beyond Regulation.* UNRISD Social Policy in a Development Context Series. New York: Palgrave Macmillan.

Rais, Jorge Carlos, Maria Esther Esquivel, and Sergio Sour. 2002. *La Concesión de los servicios de agua potable y alcantarillado sanitario en Tucumán, República Argentina.* Washington, DC: PPIAF.

Renzetti, Steven. 1999. "Municipal Water Supply and Sewage Treatment: Costs, Prices and Distortions." *Canadian Journal of Economics* 32: 688–704.

Richard, B., and T. Triche. 1994. "Reducing Regulatory Barriers to Private Sector Participation in Latin America's Water and Sanitation Services." Policy Research Working Paper 1322, World Bank, Washington, DC.

Ringskog, Klas, M. Hammond, and A. Locussol. 2006. *The Impact from Management and Lease/Affermage Contracts.* Draft Report, June. Washington, DC: PPIAF and World Bank.

Rivera, D. 1996. *Private Sector Participation in the Water Supply and Wastewater Sector: Lessons from Six Developing Countries.* Directions in Development Series. Washington, DC: World Bank.

Roda, Pablo. 2003. *Análisis de la concesión de acueducto y alcantarillado en la ciudad de Montería, Colombia.* Washington, DC: Inter-American Development Bank.

Rossi de Oliveira, André. 2008. "Social Policies and Private Sector Participation in Water Supply: The Case of Brazil." In *Social Policies and Private Sector Participation in Water Supply: Beyond Regulation*, ed. Naren Prasad, 126–48. New York: Palgrave Macmillan.

Roth, Gabriel. 1987. *The Private Provision of Public Services in Developing Countries.* New York: Oxford University Press.

Saghir, Jamal, Elizabeth Sherwood, and Andrew Macoun. 1998. "Management Contracts in Water and Sanitation: The Gaza's Experience." Public Policy for the Private Sector Series Note 1777, World Bank, Washington, DC.

Saltiel, Gustavo. 2006. "La participación del sector privado en los servicios de agua y saneamiento de la provincia de Salta, Argentina." Mimeo, November, World Bank, Washington, DC.

Saussier, S. 2004. *Public-Private Partnerships and Prices: Evidence from Water Distribution in France.* Working paper, University of Paris I Sorbonne.

Saussier, Stéphane, and Claude Ménard. 2002. "Contractual Choice and Performance: The Case of Water Supply in France." In *Economics of Contracts: Theory and Applications,* ed. Eric Brousseau and Jean-Michel Glachant, 440–63. Cambridge, U.K.: Cambridge University Press.

Savedoff, William, and Pablo Spiller. 1999. *Spilled Water: Institutional Commitment in the Provision of Water Services.* Washington, DC: Inter-American Development Bank.

Sawkins, John, and Valerie Dickie. 2008. "Case of Great Britain." In *Social Policies and Private Sector Participation in Water Supply: Beyond Regulation*, ed. Naren Prasad, 70–102. New York: Palgrave Macmillan.

Schur, Michael, Stephan Von Klaudy, and Georgina Dellecha. 2006. "The Role of Developing Country Firms in Infrastructure." Gridlines 3, PPIAF, Washington, DC.

Schusterman, R., F. Almansi, A. Hardoy, G. McGranahan, I. Oliverio, R. Rozensztejn, and G. Urquiza. 2002. *Experiences with Water Provision in Four Low-Income Barrios in Buenos Aires.* London: International Institute for Environment and Development.

SDC (Swiss Agency for Development and Cooperation). 2005. *Public-Private Partnerships for Water Supply and Sanitation, Policy Principles and Implementation Guidelines for Sustainable Services.* Berne, Switzerland: SDC.

Serão da Motta, Ronaldo, and Ajax Moreira. 2004. "Efficiency and Regulation in the Sanitation Sector in Brazil." Discussion Paper 1059, Instituto de Pesquisa Econômica Aplicada, Rio de Janeiro, Brazil.

Shirley, Mary, ed. 2002. *Thirsting for Efficiency: The Economics and Politics of Urban Water System Reform.* Oxford, U.K.: Pergamon Press.

Shirley, Mary, and Patrick Walsh. 2000. "Public Versus Private Ownership: The Current State of the Debate." Policy Research Working Paper 2420, World Bank, Washington, DC.

Shirley, Mary, Colin Xu, and Ana Maria Zuluaga. 2000. "Reforming the Urban Water System in Santiago, Chile." Policy Research Working Paper 2294, World Bank, Washington, DC.

Silva Salamanca, Julio Miguel. 2007. "Reformas Estructurales en el sector de agua potable y saneamiento básico en Colombia, 1990–2006." Mimeo, World Bank, Washington, DC.

Singh, Avjeet. 2006. "Private Sector Participation in Urban Water Supply and Sanitation Sector: A Review of Empirical Evidence." Mimeo, World Bank, Washington, DC.

———. 2007. "Political Economy of Reform in the Urban WSS Sector in India: Lessons from the Delhi Water Sector Experience." Master's thesis, John F. Kennedy School of Government, Harvard University, Cambridge, MA.

Sirtaine, Sophie, Maria Elena Pinglo, J. Luis Guasch, and Vivien Foster. 2004. *How Profitable Are Infrastructure Concessions in Latin America?* Washington, DC: World Bank.

Slattery, Kathleen. 2003. *What Went Wrong: Lessons from Cochabamba, Manila, Buenos Aires, and Atlanta.* Washington, DC: Institute for Public-Private Partnerships.

Solanes, Miguel. 2006. "Efficiency, Equity, and Liberalisation of Water Services in Buenos Aires, Argentina." *Industry, Services & Trade* (22): 124–48.

Solanes, Miguel, and Andrei Jouravlev. 2007. *Revisiting Privatization, Foreign Investment, International Arbitration, and Water.* CEPAL Series 129. Santiago, Chile: CEPAL.

Solo, T. M. 2003. *Independent Water Entrepreneurs in Latin America: The Other Private Sector in Water Services.* Washington, DC: World Bank.

Stiggers, David. 1999. "Private Participation in Water and Wastewater Services in Trinidad and Tobago." In *Can Privatization Deliver?* ed. Federico Basañes and others, 245–54. Washington, DC: Inter-American Development Bank.

Torres de Mästle, Clemencia, and Ada Karina Izaguirre. 2008. "Recent Trends in Private Activity in Infrastructure: What the Shift Away from Risk Means for Policy." Gridlines 31, PPIAF, Washington, DC.

Transparency International. 2008. *Corruption in the Water Sector, Global Corruption Report 2008.* Cambridge, U.K.: Cambridge University Press.

Tremolet, Sophie, and Catherine Hunt. 2006. "Taking Account of the Poor in Water Sector Regulation." Water Supply and Sanitation Working Note 11, World Bank, Washington, DC.

Triche, T., A. Mejia, and E. Idelovitch. 1993. "Arranging Concessions for Water Supply and Sewerage Services. Lessons from Buenos Aires and Caracas." Infrastructure Note 10, World Bank, Washington, DC.

Triche, Thelma, Sixto Requena, and Mukami Kariuki. 2006. "Engaging Local Private Operators in Water Supply and Sanitation Services, Initial Lessons from Emerging Experiences in Cambodia, Colombia, Paraguay, the Philippines, and Uganda." Water Supply and Sanitation Working Note 12, World Bank, Washington, DC.

Urquhart, Penny, and Deborah Moore. 2004. *Global Water Scoping Process: Is There a Case for a Multistakeholder Review of Private Sector Participation in Water and Sanitation?* London: ASSEMAE (Brazilian Association of Municipal Water and Sanitation Public Operators), Consumers International, Environmental Monitoring Group, Public Services International, RWE-Thames Water, and WaterAid. http://www.wateraid.org.

Vagliasindi, Maria, and Ada Karina Izaguirre. 2007. "Private Participation in Infrastructure in Europe and Central Asia: A Look at Recent Trends." Gridlines 26, PPIAF, Washington, DC.

Valfrey-Visser, Bruno, David Schaub-Jones, Bernard Collignon, and Emmanuel Chapponière. 2006. *Access through Innovation: Expanding Water Services Delivery through Independent Network Providers.* London: Building Partnerships for Development; and Paris: AFD (French Development Agency).

Van den Berg, Caroline. 1999. "Water Privatization in England and Wales." Viewpoint Note 115, World Bank, Washington, DC.

———. 2000. "Who Wins, Who Loses, and What To Do About It." Viewpoint Note 217, World Bank, Washington, DC.

Van den Berg, Caroline, S. Pattanayak, C. J. C. Yang, and Gunatilake Herath. 2006. "Getting the Assumptions Right: Private Sector Participation Transaction Design and the Poor in Southwest Sri Lanka." Water Sector Board Discussion Paper 7, World Bank, Washington, DC.

Van Ginneken, Meike, Ross Tyler, and David Tagg. 2004. "Can the Principles of Franchising Be Used to Improve Water and Sanitation Services? A Preliminary Analysis." Water Supply and Sanitation Sector Board Discussion Paper Series 2, World Bank, Washington, DC.

Veevers-Carter, Patricia. 2005. "Emerging Applications in PPP in Water Supply and Sanitation." Presentation at Water Week, World Bank, Washington, DC, March.

Vives, Antonio, Angela Paris, and Juan Benavides. 2006. *Financial Structuring of Infrastructure Projects in PPP: An Application to Water Projects.* Washington, DC: Inter-American Development Bank.

Walton, B., and M. Sohail. 2003. *Public Private Partnerships and the Poor: Bolivia—A Perspective on Water Supply and Sewerage.* Leicestershire, U.K.: WEDC, Loughborough University.

WaterAid and Tearfund. 2003. *Advocacy Guide to Private Sector Involvement in Water Services.* London: WaterAid and Tearfund.

Wenyon, Sylvia, and Charles Jenne. 1999. "Water and Sewerage Privatization and Reforms." In *Can Privatization Deliver?* ed. Federico Basañes and others, 73–215. Washington, DC: Inter-American Development Bank.

WHO/UNICEF MDG Joint Monitoring Program. http://www.wssinfo.org.

Winpenny, Jim. 2006. *Opportunities and Challenges Arising from the Increasing Role of New Private Operators in Developing and Emerging Economies.* Paris: Global Forum on Sustainable Development.

Wolff, G., J. Moss, G. Gladden, and E. Guttierrez. 2003. "Valuing Water for Better Governance." Business and Industry CEO Panel for Water White Paper, Pacific Institute, Oakland, CA.

Wolff, G., and M. Palaniappan. 2004. "Public or Private Water Management? Cutting the Gordian Knot." *Journal of Water Resources Planning and Management* 3 (February).

World Bank. 1995. *Bureaucrats in Business: The Economics and Politics of Government Ownership.* World Bank Policy Research Report. New York and London: Oxford University Press.

———. 2004a. "Morocco: Recent Developments in Infrastructure, Water Supply, and Sanitation." Report 29634–MOR, World Bank, Washington, DC.

———. 2004b. *Public and Private Sector Roles in Water Supply and Sanitation Services: Operational Guidance for World Bank Group Staff.* Washington, DC: World Bank.

———. 2006. *Approaches to Private Sector Participation in Water Services: A Toolkit.* Washington, DC: World Bank and PPIAF.

WSP (Water and Sanitation Program). 2001. *The Buenos Aires Concession: The Private Sector Serving the Poor.* WSP South Asia Series, Washington, DC: World Bank.

———. 2006. "Jamshedpur Utilities and Services Company Limited: Improving WSS Services through Private Sector Partnerships." WSP field note, World Bank, Washington, DC.

Wu, Xun, and Nepomuceno Malaluan. 2008. "A Tale of Two Concessionaires: A Natural Experiment of Water Privatisation in Metro Manila." *Urban Studies Journal* 45 (1): 207–29.

Yahya, Jamal, and Leroy Jones. 2006. *Impact of Privatization in Africa: Senegal Water.* Boston: Boston Institute for Developing Economies.

———. 2006b. *Impact of Privatization in Africa: Uganda Water.* Boston: Boston Institute for Developing Economies.

Yepes, Guillermo. 2007. "PPP for Urban WSS Services in Latin America: Three Case Studies in Argentina." Mimeo, May, World Bank, Washington, DC.

Zerah, Marie Hélène. 2000. *The Cancellation of the Pune Water Supply and Sewerage Project.* Water and Sanitation Program (WSP) South Asia Series. Washington, DC: World Bank.

# INDEX

Boxes, figures, notes, and tables are indicated by *b*, *f*, *n*, and *t*, respectively.

## A

## C

## H



## I



## J



## K

## L



## M

## N



## O

## T

**Eco-Audit**

*Environmental Benefits Statement*

The World Bank is committed to preserving endangered forests and natural resources. The Office of the Publisher has chosen to print *Public-Private Partnerships for Urban Water Utilities* on recycled paper with 30 percent postconsumer fiber in accordance with the recommended standards for paper usage set by the Green Press Initiative, a nonprofit program supporting publishers in using fiber that is not sourced from endangered forests. For more information, visit www.greenpressinitiative.org.

Saved:

- 13 trees
- 4 million British thermal units of total energy
- 1,218 pounds of net greenhouse gases
- 5,864 gallons of waste water
- 356 pounds of solid waste